# Air Quality Management

# ISSUES IN ENVIRONMENTAL SCIENCE AND TECHNOLOGY



TITLES IN THE SERIES:

1 Mining and its Environmental Impact
2 Waste Incineration and the Environment
3 Waste Treatment and Disposal
4 Volatile Organic Compounds in the Atmosphere
5 Agricultural Chemicals and the Environment
6 Chlorinated Organic Micropollutants
7 Contaminated Land and its Reclamation
8 Air Quality Management

FORTHCOMING:

9 Risk Assessment and Management
10 Air Pollution and Health

*How to obtain future titles on publication*

A subscription is available for this series. This will bring delivery of each new volume immediately upon publication. For further information, please write to:

The Royal Society of Chemistry
Turpin Distribution Services Limited
Blackhorse Road
Letchworth
Herts SG6 1HN, UK

Telephone: +44 (0) 1462 672555
Fax: +44 (0) 1462 480947

ISSUES IN ENVIRONMENTAL SCIENCE AND TECHNOLOGY

EDITORS: R.E. HESTER AND R.M. HARRISON

# 8
# Air Quality Management

THE ROYAL SOCIETY OF CHEMISTRY
Information Services

ISBN 0-85404-235-0
ISSN 1350-7583

A catalogue record for this book is available from the British Library



Published by The Royal Society of Chemistry, Thomas Graham House,
Science Park, Milton Road, Cambridge CB4 4WF, UK

Typeset in Great Britain by Vision Typesetting, Manchester
Printed and bound by Redwood Books Ltd., Trowbridge, Wiltshire

# Preface

Concern over air quality has never been higher in the public mind in both developed and less developed countries. This concern is reflected in ever tightening legislation, and the vigorous activity of regulatory authorities. The easily won gains in air quality have already been made throughout the developed world by the implementation of inexpensive but effective control measures. Air quality management is therefore addressing an ever steepening part of the cost/benefit curve, whereby each incremental improvement in air quality becomes increasingly expensive as the atmosphere becomes cleaner. Thus, methods to quantify the improvements required, to predict the source controls most appropriately applied, and to provide cost/benefit analyses of the reductions are becoming increasingly sophisticated. This volume deals with the scientific aspects of such air quality management procedures.

The first article, by D. R. Middleton, sets the scene by describing from a UK perspective the developments currently in hand to provide a scientific basis for air quality management. Subsequent articles deal with specific components of the air quality management process. Effective control of primary air pollution depends critically upon good knowledge of the sources of emissions and their geographic locations. This is encompassed by emissions inventories, and the article by D. Hutchinson deals with the now rather sophisticated subject of compilation of source emissions inventory data; it is illustrated by reference to the recently compiled emissions inventory for the UK West Midlands. A second crucial aspect of air quality management is the monitoring of air pollution. This provides information on the temporal trends in air quality and gives a direct measure of the compliance or otherwise of air pollutant concentrations with ambient air quality standards designed to protect human health, ecosystem function or the integrity of inert materials. The design and operation of air monitoring networks is described in the third article by J. Bower. Both source emissions data and monitoring information are central to the activity described in the fourth article by A. Skouloudis, who gives a comprehensive overview of the scientific considerations in the European Auto-Oil study, which was sponsored by the European Commission in order to determine the most cost-effective means of meeting air quality targets. It was therefore central to the setting of vehicle emission and fuel quality standards for implementation in the years 2000 and 2005.

Secondary air pollutants are those formed within the atmosphere and these present considerable difficulties in evaluating the effectiveness of controls of precursor emissions upon concentrations of the secondary pollutant. Often the relationship between emission of the precursor and concentration of the pollutant are strongly non-linear. The fifth article by P. Hopke on source–receptor modelling of air pollution addresses the very difficult technical issue of how, starting from ambient air quality data, it is possible to identify the source areas contributing to secondary pollutants to the atmosphere. The sixth article by M. Hornung and colleagues deals with the use of geochemical or biological tolerances (critical loads) to determine the maximum allowable inputs of acidic pollutants to the terrestrial environment. This approach is now heavily embodied in air pollution control policy development within Europe. The final article by A. C. Lloyd gives a fascinating perspective of the successes and problems of air pollution control in California, the US state which has for many years led the way in promoting vigorous air pollution control measures, but still has massive problems to overcome.

We are very fortunate in having attracted articles from leading workers in this field representing the best of scientific endeavour from both Europe and North America. We are most grateful to them for providing readers with a comprehensive perspective of the current state of the art of air quality management.

Roy M. Harrison
Ronald E. Hester

# Contents

# Editors

**Ronald E. Hester, BSc, DSc(London), PhD(Cornell), FRSC, CChem**

Ronald E. Hester is Professor of Chemistry in the University of York. He was for short periods a research fellow in Cambridge and an assistant professor at Cornell before being appointed to a lectureship in chemistry in York in 1965. He has been a full professor in York since 1983. His more than 300 publications are mainly in the area of vibrational spectroscopy, latterly focusing on time-resolved studies of photoreaction intermediates and on biomolecular systems in solution. He is active in environmental chemistry and is a founder member and former chairman of the Environment Group of The Royal Society of Chemistry and editor of 'Industry and the Environment in Perspective' (RSC, 1983) and 'Understanding Our Environment' (RSC, 1986). As a member of the Council of the UK Science and Engineering Research Council and several of its sub-committees, panels and boards, he has been heavily involved in national science policy and administration. He was, from 1991–93, a member of the UK Department of the Environment Advisory Committee on Hazardous Substances and is currently a member of the Publications and Information Board of The Royal Society of Chemistry.

**Roy M. Harrison, BSc, PhD, DSc (Birmingham), FRSC, CChem, FRMetS, FRSH**

Roy M. Harrison is Queen Elizabeth II Birmingham Centenary Professor of Environmental Health in the University of Birmingham. He was previously Lecturer in Environmental Sciences at the University of Lancaster and Reader and Director of the Institute of Aerosol Science at the University of Essex. His more than 250 publications are mainly in the field of environmental chemistry, although his current work includes studies of human health impacts of atmospheric pollutants as well as research into the chemistry of pollution phenomena. He is a former member and past Chairman of the Environment Group of The Royal Society of Chemistry for whom he has edited 'Pollution: Causes, Effects and Control' (RSC, 1983; Third Edition, 1996) and 'Understanding our Environment: An Introduction to Environmental Chemistry and Pollution' (RSC, Second Edition, 1992). He has a close interest in scientific and policy aspects of air pollution, currently being Chairman of the Department of Environment Quality of Urban Air Review Group as well as a member of the DoE Expert Panel on Air Quality Standards and Photochemical Oxidants Review Group and the Department of Health Committee on the Medical Effects of Air Pollutants.

# Contributors

**J. Bower,** *AEA Technology, National Environmental Technology Centre, Culham, Oxfordshire OX14 3DB, UK*

**H. Dyke,** *Institute of Terrestrial Ecology, Monks Wood Research Station, Abbots Ripton, Huntingdon, Cambridgeshire PE17 2LS, UK*

**J. R. Hall,** *Institute of Terrestrial Ecology, Monks Wood Research Station, Abbots Ripton, Huntingdon, Cambridgeshire PE17 2LS, UK*

**P. K. Hopke,** *Department of Chemistry, Clarkson University, Box 5810, Potsdam, New York 13699, USA*

**M. Hornung,** *Institute of Terrestrial Ecology, Merlewood Research Station, Grange over Sands, Cumbria LA11 6JU, UK*

**D. Hutchinson,** *London Research Centre, 81 Black Prince Road, London SE1 7SZ, UK*

**A. C. Lloyd,** *Energy and Environmental Engineering Center, Desert Research Institute, PO Box 60220, Reno, Nevada 89506, USA*

**S. E. Metcalfe,** *Department of Geography, University of Edinburgh, Edinburgh EH8 9XP, UK*

**D. R. Middleton,** *Meteorological Office, London Road, Bracknell, Berkshire RG12 2SZ, UK*

**A. N. Skouloudis,** *European Commission, DG Joint Research Centre 'Ispra', Environment Institute, TP 250, Ispra (VA), I-21020, Italy*

# Improving Air Quality in the United Kingdom

DOUGLAS R. MIDDLETON*

## 1 Introduction

The concentration of pollutants in urban areas from sources near the ground has become of increasing concern in the UK, particularly since the London pollution episode of December 1991. During this episode from 11 to 15 December 1991, $NO_2$ was unusually high, with values from 350 to 400 ppb recorded at several sites and reaching 423 ppb at Bridge Place.[1,2] These values were well above the standard[3] which recommended a maximum hourly average concentration for $NO_2$ of 150 ppb. A study[4] by South East Institute of Public Health mapped contours of air pollution measurements in the form of annual average concentrations of nitrogen dioxide ($NO_2$). During 1995, these contours in much of London were above the $NO_2$ Standard that appears in the UK National Air Quality Strategy.[1] This Standard is 21 ppb for an annual mean. Since in London some 76% of the emissions of oxides of nitrogen are associated with road transport,[5] measures for improving air quality will have to address transport planning. Looking forward, the report[2] shows a projection of $NO_2$ contours in London for the year 2000. It suggests that by that date in central London, $NO_2$ will still be likely to exceed $40\,\mu g\,m^{-3}$ (*i.e.* 21 ppb).

The Environment Act 1995 has increased the powers for local authorities to manage air quality and to consult with a wide range of bodies. The decisions to be taken in managing air quality will cross traditional boundaries. It will be necessary, for example, for Environmental Health Officers and environmental scientists in local government to liaise quite closely with traffic planners and

*Any views expressed are those of the author.



[1] DoE, *The United Kingdom National Air Quality Strategy*, Department of the Environment CM 3587, HMSO, London, 1997.

[2] QUARG, *Urban Air Quality in the United Kingdom*, first report of the Quality of Urban Air Review Group, Department of the Environment, London, 1993.

[3] EPAQS, *Expert Panel on Air Quality Standards* (*Nitrogen Dioxide, Ozone, Carbon Monoxide, Sulphur Dioxide, Benzene, 1,3-Butadiene*), Department of the Environment, HMSO, London, 1996.

[4] SEIPH, *Air Quality in London 1995*, third report of the London Air Quality Network, South East Institute of Public Health, Tunbridge Wells, Kent, 1996.

[5] M. Chell and D. Hutchinson, *London Energy Study*, London Research Centre, London, 1993.

highway engineers, as well as with managers of local industry and the Environment Agency inspectors, in order to assemble the emissions databases upon which the modelling of air quality for reviews and assessments will rely. The National Strategy for Air Quality[1] sets out Government policy with regard to improving ambient air quality in the UK. It looks to the year 2005 and is relevant to both statutory requirements and to further voluntary action.

The Environment Act 1995 requires Local Authorities to review past and assess future air quality within their areas of jurisdiction. It also requires them to identify areas where levels of pollutants are high and, if necessary, designate them to be local air quality management areas. Such areas will be defined using the Standards and Objectives which appear in Table 3.1 of the Strategy.[1] Management areas might be based upon administrative or other boundaries, but will need to contain locales where air quality will be high to the year 2005. Computer models provide not only a means of forecasting air pollution events, but also ways to investigate the contribution of actual or planned pollution sources. Similarly, monitoring data can be used as a means of projecting forward in order to assess the extent of control that is needed. Monitoring data are costly to acquire, and do not easily lend themselves to consideration of future controls on individual sources. There will therefore be more use of emissions inventories and modelling to complement the monitoring of pollutants.

We have here an example of how the new moves to improve air quality through local management are providing a stimulus to scientific research. As well as encouraging the development of simple assessment techniques, it has led to the need to develop simple monitoring methods for pollutants such as buta-1,3-diene, along the lines of the diffusion tubes for benzene. There is also work now in progress by a number of local authority groupings funded by the Department of the Environment (DoE) to validate dispersion models against each other and against measurements. Finally, we shall see below that the UK National Air Quality Strategy[1] presents Standards and Objectives that will serve as a yardstick against which to judge air quality improvements.

## 2 Pollution in Street Canyons

Much monitoring has focused on the background level of airborne substances such as Pb, $NO_x$, CO, $O_3$, particles and organic compounds.[2,6–8] However, it is at kerbside locations where the general public may suffer exposures to the highest concentrations of pollutants. This is particularly true in a street canyon,[9] a

[6] QUARG, *Vehicle Emissions and Urban Air Quality*, second report of the Quality of Urban Air Review Group, Institute of Public and Environmental Health, University of Birmingham, Birmingham, 1993.

[7] QUARG, *Airborne Particulate Matter in the United Kingdom*, third report of the Quality of Urban Air Review Group, Institute of Public and Environmental Health, University of Birmingham, Birmingham, 1996.

[8] V. Bertorelli and R. G. Derwent, *Air Quality A to Z: A Directory of Air Quality Data for the United Kingdom in the 1990s*, Meteorological Office, Bracknell, Berks, 1995.

[9] A. T. Buckland, *Validation of a Street Canyon Model in Two Cities*, draft paper: IOP conference on 'Urban Air Quality—Monitoring and Modelling', Hatfield, 11–12 July 1996, *Environ. Monitoring Assessment*, 1997, submitted.

relatively narrow street between buildings that line up continuously along both sides. The combination of large vehicle emissions and reduced dispersion in these circumstances can lead to high levels of pollution. A well-trafficked street canyon therefore represents an important facet of air quality management.[9]

Recognizing that some authorities may not have the resources to run elaborate models, but will need a quantitative method, Buckland and Middleton[10] have produced a simple method. The result is AEOLIUS, a selection of nomograms and charts that has been devised along similar lines to volume 11 of the Design Manual for Roads and Bridges.[11] AEOLIUS is designed with one purpose in mind: to estimate the likely maximum concentrations from traffic in urban street canyons. It does not include the additional background concentrations from sources outside the street; they should be added by the user as necessary. AEOLIUS and other models are being tested by some local authorities during the trials cited earlier.

Only a brief outline of canyon dispersion principles is made here. A fuller description of canyon models appears in Buckland[9] and the papers cited therein. When the wind blows across a street canyon a vortex is typically generated, with the wind flow at street level opposite to that above roof level. A consequence is lower concentrations of pollutants on the windward side of the street compared with those of the leeward. The windward side is here defined as the side the roof wind blows to whilst the leeward side is the side the roof wind blows from. The quantity of pollutant that a monitor directly receives from vehicle emissions is calculated using a simple Gaussian model. The contribution from air recirculated by the vortex is calculated using a simple box model. The principle is that the inflow rate of pollutant into the volume of recirculated air is equal to the outflow rate and that the pollutants are well mixed inside this volume.

The canyon concentration[9] is proportional to the total emission rate $Q$ from all vehicles, which will reflect changes in the vehicle fleet emission factors. For $N$ vehicles an hour (with all vehicle types combined) and a combined emission factor of $q$ grams km$^{-1}$ vehicle$^{-1}$, the total emision rate $Q$ $\mu$g m$^{-1}$ s$^{-1}$ is given by the equation: $Q = Nq/3.6$, which converts grams to micrograms, km$^{-1}$ to m$^{-1}$ and vehicles hr$^{-1}$ to vehicles s$^{-1}$. This equation means that in order to improve air quality by reducing motor vehicle emissions $Q$, it is necessary to reduce the number of vehicles $N$ and their average mass emission factor $q$. When $Q$ is used in a dispersion model to calculate air pollutant concentration, it is multiplied by road length or distance travelled $S$; to improve air quality this also indicates reductions in the distance $S$. Traffic management and public transport can be regarded as managing $q$, $N$ and the distance travelled $S$, whilst changes in technology such as catalysts or particle traps on new vehicles, and the maintenance of existing vehicles, seek to reduce $q$. Other developments such as cleaner diesel or unleaded petrol also serve to reduce $q$ (for particulates or lead, respectively). Similar principles apply to fast moving traffic on open roads, and to idling engines in congestion. When vehicles are first started, and the catalyst is

[10] A. T. Buckland and D. R. Middleton, *Nomograms for Street Canyons*, 1997, submitted to *Atmos. Environ.*

[11] TRL, *Design Manual for Roads and Bridges*, vol. 11, sect. 3, part 1—Air Quality, Transport Research Laboratory, Department of Transport, amended 1994, HMSO, London, 1995.

**Table 1** Road transport emission factors[12] $q$ for urban traffic flow with fleet averages[10]

| *Vehicle type* | *Number of vehicles per hour* | $NO_x$/g km$^{-1}$ | CO/g km$^{-1}$ |
|---|---|---|---|
| Cars* | 830 | 2.11 | 19.36 |
| LGV* | 80 | 1.49 | 11.20 |
| Medium HGV | 40 | 12.60 | 6.00 |
| Large HGV | 10 | 16.95 | 7.30 |
| Bus | 10 | 14.40 | 6.60 |
| Motorcycle | 30 | 0.30 | 20.00 |
| All types | 1000 | 2.70 | 17.94 |

*When calculating the factors for an average car or light goods vehicle, the assumed percentages of petrol and diesel were: cars 90%P, 10%D; light goods vehicles 50%P, 50%D.

cold, the value of $q$ may be much larger than for normal driving. Some published values[12] for $q$ are shown in Table 1; a busy traffic flow $N$ might be 2000 vehicles per hour. Finally, it can be seen from the Strategy[1] that even as average $q$ is reduced in coming years, growths in $N$ and $S$ beyond the year 2010 are projected to outweigh the benefits of catalysts. The Strategy,[1] p. 46, therefore lists the principles for improving air quality, which we summarize here:

- improved technology
- tighter fleet management
- environmentally responsible use of vehicles
- policies and planning to reduce reliance on cars.

## 3 Motor Vehicle Contribution

The importance of motor vehicles to the urban air quality debate is manifest in figures published by the newly completed West Midlands Emissions Inventory study. This survey of all major sources of air pollutants in the seven local authorities was funded by the DoE to provide the first inventory in the UK at a very detailed local scale. Hutchinson and Clewley[13] collected data on transport (road, rail, air), domestic and industrial, Part B (*i.e.* those under local authority control), and Part A (*i.e.* under Environment Agency control), processes. Emissions of sulfur dioxide, oxides of nitrogen, carbon monoxide, carbon dioxide, non-methane volatile organic compounds, benzene, buta-1,3-diene and particles were mapped on 1 km squares. The database contains approximately 9000 road links (derived from the data used to run the traffic model for the West Midlands) with their start and end co-ordinates, peak hour traffic flows and vehicle speeds. They[13] concluded that:

> 'The single most significant source of atmospheric pollutants in the West Midlands is road traffic. In the case of carbon monoxide, benzene and

[12] C. A. Gillham, P. K. Leech and H. S. Eggleston, *UK Emissions of Air Pollutants 1970–1990*, DoE Document LR 887 (AP), Department of the Environment, London, 1992.

[13] D. Hutchinson and L. Clewley, *West Midlands Atmospheric Emissions Inventory*, London Research Centre, London, 1996.

**Table 2** Road transport emission factors representing the national vehicle fleet for the year 1996

| *Vehicle type* | *Number of vehicles per hour* | $NO_x$/ g km$^{-1}$ | CO/ g km$^{-1}$ | $PM_{10}$/ g km$^{-1}$ | $SO_2$/ g km$^{-1}$ | *Benzene*/ g km$^{-1}$ | *Buta-1,3-diene*/ g km$^{-1}$ |
|---|---|---|---|---|---|---|---|
| Cars P | 747 | 1.57 | 14.86 | 0.028 | 0.051 | 0.0917 | 0.0267 |
| Cars D | 83 | 0.57 | 0.19 | 0.156 | 0.059 | 0.0042 | 0.0068 |
| LGV P | 40 | 1.51 | 18.37 | 0.044 | 0.168 | 0.1499 | 0.0349 |
| LGV D | 40 | 1.41 | 1.00 | 0.259 | 0.059 | 0.0065 | 0.0104 |
| Medium HGV | 40 | 6.87 | 4.99 | 1.187 | 0.229 | 0.0106 | 0.0136 |
| Large HGV | 10 | 13.17 | 5.99 | 0.996 | 0.387 | 0.0106 | 0.0220 |
| Bus | 10 | 14.75 | 18.73 | 1.347 | 0.352 | 0.0108 | 0.0198 |
| Motorcycle | 30 | 0.20 | 16.53 | 0.087 | 0.023 | 0.0404 | 0.0148 |
| All types | 1000 | 1.897 | 12.83 | 0.1195 | 0.0693 | 0.07695 | 0.02373 |

Source: T. Murrells, NETCEN AEA Technology (personal communication, see Buckland and Middleton[10]).

**Table 3** Some scaling factors to reflect changing emissions 1996–2005*

| *Year* | $NO_x$ | CO | $PM_{10}$ | $C_6H_6$ | $C_4H_6$ |
|---|---|---|---|---|---|
| 1996† | 1.0 | 1.0 | 1.0 | 1.0 | 1.0 |
| 1997 | 0.905 | 0.889 | 0.827 | 0.895 | 0.874 |
| 1998 | 0.824 | 0.813 | 0.717 | 0.800 | 0.763 |
| 1999 | 0.752 | 0.763 | 0.626 | 0.715 | 0.668 |
| 2000 | 0.688 | 0.698 | 0.539 | 0.636 | 0.574 |
| 2001 | 0.626 | 0.636 | 0.484 | 0.560 | 0.489 |
| 2002 | 0.570 | 0.580 | 0.433 | 0.491 | 0.411 |
| 2003 | 0.517 | 0.527 | 0.402 | 0.427 | 0.335 |
| 2004 | 0.489 | 0.491 | 0.381 | 0.382 | 0.287 |
| 2005 | 0.455 | 0.461 | 0.358 | 0.344 | 0.243 |

*Values were based on national vehicle fleet estimates of emission factors (kindly supplied by NETCEN AEA Technology) for each vehicle type, and weighted here by the traffic pattern of Table 1 as discussed in the text (see also Buckland and Middleton[10]).
†Values in each year are normalized by the last row of Table 2, the 1996 revised emission factors for an average vehicle.

> buta-1,3-diene, road traffic accounts for over 96% of emissions. . . . Road traffic also accounts for 85% of emissions of oxides of nitrogen and 75% of black smoke, but only 16% of sulfur dioxide.'

It is therefore clear that improvements in air quality will be dependent upon measures to reduce the total rate of emissions ($SNq$) from the motor vehicle fleet.

The composition of the vehicle fleet in terms of catalysts and type of fuel will change. To reflect this, revised emission factors (Table 2) provided by NETCEN AEA Technology for the expected national fleet in future years were also used[10] to estimate the likely fractional changes in $q$ (Table 3). The decrease in revised emission factor for each pollutant from 1996 to 2005 is in Table 3 where the values shown are normalized to the 1996 estimates of revised emissions, *e.g.* for a CO calculation in the year 2001, the emissions must be multiplied by 0.636. In the absence of detailed local traffic analyses in different towns, Table 3 should provide reasonable estimates of future trends in vehicle emissions. Similar information on emission factors and their future trends appears in the Design Manual for Roads and Bridges,[11] and in the Strategy.[1] Finally, whilst on the subject of models, Middleton[14] reviewed physically based models for use in air quality management, for local or distant plumes.

## 4 Future Air Quality Objectives

In order to improve air quality, the targets at which to aim must be made visible. The Strategy[1] gives Standards for priority pollutants and possible Objectives for achieving them. Table 4 is derived from the Strategy and EPAQS[3] reports. Compliance or otherwise with the proposed values will be assessed for the year 2005. Breaching of these may point to a detailed assessment being necessary and

[14] D. R. Middleton, *Physical Models of Air Pollution for Air Quality Reviews*, Clean Air 26 (2), National Society for Clean Air and Environmental Protection, Brighton, 1996, pp. 28–36.

**Table 4** Air quality quantities

| | |
|---|---|
| Benzene | Strategy 5 ppb running annual mean; EPAQS[3] also recommended reduction to 1 ppb; timescale not specified |
| Buta-1,3-diene | Strategy 1 ppb running annual mean; EPAQS[3] recommended review in five years |
| Carbon monoxide | Strategy 10 ppm running 8 hour mean; as EPAQS[3] |
| Lead | Strategy 0.5 $\mu g\,m^{-3}$ annual mean; EPAQS (pending) lower? |
| Nitrogen dioxide | Strategy 150 ppb 1-hour mean; as EPAQS[3] Strategy also 21 ppb annual mean |
| Ozone* | Strategy 50 ppb running 8 hour mean; as in EPAQS[3] Objective 97th percentile of running 8 hourly means not to exceed standard value above of 50 ppb |
| Fine particles $PM_{10}$ | Strategy 50 $\mu g\,m^{-3}$ running 24 hour mean; as in EPAQS[3] Objective 99th percentile of running 24 hourly means not to exceed standard value above of 50 $\mu g\,m^{-3}$ |
| Sulfur dioxide | Strategy 100 ppb as 15 minute mean; as in EPAQS[3] Objective 99.9th percentile of 15 minute means not to exceed standard value above of 100 ppb |

Source: the National Strategy[1] and others.[3]
*Ozone is a pollutant formed by chemical reactions in the atmosphere taking place over large distances; national and international measures are seen as the likely approach.[1]

the likely declaration of an Air Quality Management Area. This will in turn invoke the need for a formal statement or Action Plan; in this the local authority will set out its programme of local action, including the pollution controls as needed to achieve compliance with the Objectives. Such plans may estimate the amount of future emission control that is indicated, and imply constraints on planning.

To identify pollution 'hot spots', air quality management areas, calculations are likely to be required at many receptors so that mapping can be carried out. Table 3.1 of the Strategy[1] has 15 minute mean, 1 hour mean, running 8 hour mean, running 24 hour mean, annual mean and running annual mean. The list also uses percentiles of some of these quantities.

Some quantities are to be used as their means for testing likely compliance by 2005 AD, *i.e.* for benzene, buta-1,3-diene, carbon monoxide, lead and nitrogen dioxide. Others involve their percentiles (on percentiles, see Spiegel[15]; on log-normal and Weibull distributions to describe air quality data, see Seinfeld[16]). It is then their concentrations at the various percentiles that are tested for compliance. They include 97th, 99th and 99.9th percentiles, which need additional processing of model results to generate the cumulative frequency distribution and derive the relevant percentile, *i.e.* ozone (not a local authority modelling task, *cf.* the DoE[1]), particles as $PM_{10}$ and $SO_2$. After the calculations, running means and percentiles can be evaluated (using spreadsheets, or routines within some models). Surrogate

[15] M. R. Spiegel, *Probability and Statistics*, McGraw-Hill, Maidenhead, 1980.
[16] J. H. Seinfeld, *Atmospheric Chemistry and Physics of Air Pollution*, Wiley, Chichester, 1986.

statistics might also be considered, such as appear in the EPAQS standard for the maximum 15 minute means of $SO_2$ *versus* the hourly means. In general, the modelling will need to calculate hourly averages as sequential means and then apply the relevant running means. For mapping air quality and the publication of results, annual averages are often convenient, such as in SEIPH.[4] They give a clear overview of the variations in air quality across an area. Maps of the Standards and Objectives in the Strategy may need special treatment, as outlined below.

## 5 Mapping an Air Quality Management Area

The Environment Act 1995 seeks to establish improvements in air quality through the identification of air quality management areas. At the time of writing, the manner in which such areas will be defined has yet to be announced; it is likely to be part of the regulations or guidance. Nevertheless, it is important to analyse ways in which such areas might be mapped. Future planning by local authorities may be strongly influenced by the projected extent of such areas, and public decision making in this area will need to be transparent and accountable. The following suggestions give an idea of the special nature of this mapping of regions that might exceed air quality Standards or Objectives.

### *Mapping Principles*

In pollution modelling the phrase 'long term' is often used to convey the notion that the average concentration is to be obtained for some very long time period, such as one year or even 10 years. Annual average concentration is a typical quantity of this sort and would average some 8760 hourly values in a year. For reliable statistics, a 10 year run which requires 87 600 iterations might model the plumes from all sources, and generate a time series of results that would then be averaged. To map the likely pollution 'hot spots' using the annual average concentration, each hourly run would have to be repeated at each receptor position over the area to be mapped. Where the criterion for compliance to identify a problem area is based upon a concentration at some percentile, then instead of the average being found at each receptor, the results must be used to obtain the percentile at each receptor. The mapped 'hot spot' is the area enclosed within a contour that follows the positions where the percentile either exceeds or is less than the standard. For $NO_2$ a mapped area may enclose receptor points with maximum values greater than the standard 150 ppb (Table 4) and mark these as exceeding the Objective. Receptors with maximum values less than 150 ppb would be judged as complying with the Objective. Alternatively, there is scope in the Strategy[1] to map an area where annual average concentration of $NO_2$ may exceed 21 ppb.

On a map to show exceedance of the Objective for Fine Particles as $PM_{10}$, the exceedance area would be the region where the receptors have 99th percentile $c_p > 50\ \mu g\ m^{-3}$. In the case of Particles, the concentrations must be expressed in the form of running 24 hour means before the percentiles of the distribution are evaluated at each receptor.

### *Modelling*

The input data must be formatted to suit the chosen model. Although broadly the same emissions inventory information are needed to model a given stack or line source, each computer code is likely to have its own sequence of data entry and slight variations on the parameters that are needed. For example, effluents from a chimney require stack height, stack diameter, exit momentum, exit buoyancy flux and emission rate of pollutant. Some models may require details of nearby buildings, whilst others will not.

The next step will be to prepare the meteorological data file. Either a series of short-term hourly observations of meteorology will be needed, or a frequency table sorted into joint categories by wind direction, stability class and wind speed. This choice of sequential *versus* statistical data can be decided using Table 4 for each pollutant in turn.

When the emissions data file and the meteorological data file are ready, the receptor layout must be decided upon. This can range from a short list of locations, to a line or grid of receptors. If aiming to map the quantities from the UK National Air Quality Strategy as listed below, then a large number of receptors may be indicated. If so, particular care is needed in identifying the region to be mapped, as a large grid of closely spaced receptors means a large number of dispersion calculations. Near to ground level sources such as roads, it can be necessary for hourly averages to have receptors as close together as every 10 metres in order to obtain meaningful contours when near the source. For annual averages a coarser spacing may suffice. The user of the model should obtain some sensitivity studies (*cf.* Royal Meteorological Society Guidelines cited below), to check that any contouring is robust and not adversely affected by an inappropriate receptor layout. Finally, we note for running means (*cf.* Table 4) that results in time and date order will be averaged in overlapping groups. Frequency data are not appropriate when running means are required. It is expected that guidance will advise local authorities on appropriate methods to map air quality management areas.

## 6 Quality Control

Meteorological data should be observed (where practicable) to the recognized World Meteorological Organization standards, and quality assured after recording and before use. There is also a need to ensure that meteorological data are from a relevant site. For urban pollution studies, it may be necessary to use measurements taken within the urban area and this poses particular problems of instrument siting and exposure; some measurements may be unduly influenced by some upwind building or structure. However, models such as street canyon models require (in principle at least) a roof-top wind; this will rarely be available so the models will rely on a 10 metre wind speed recorded at an observing station outside the town. AEOLIUS, for example, was designed[10] to use a 10 m wind when used for screening, although the detailed calculations within the model then extrapolate this wind to roof height. Other models such as the Indic model can accept meteorological data directly from a data logger that records the signals from instruments mounted on a mast in the urban area, whereas the ADMS model was designed for flexibility in its data requirements.

When calculating percentiles and running means, adequate data capture is required. For running means it is recommended that 75% data capture (*e.g.* 6 hours out of every group of 8 consecutive hours, or 16 out of every 24) be required; if less than this is available, then the running mean is not acceptable. Running means should be stored by the time/date at the end of the period.

The Royal Meteorological Society has issued a Policy Statement[17] for the use and choice of models. Dispersion modelling may be part of a public decision making process, with the results being entered into the public record. We may expect air quality reviews and assessments to be public documents, so any dispersion modelling runs may be subject to public scrutiny. Therefore in seeking to make plans to improve air quality, just as in the regulating of industrial processes, full documentation to establish an audit trail should be prepared. The Guidelines[17] establish principles for a technically valid and properly communicated modelling exercise.

## 7 Other Approaches

### *Surrogate Statistics*

In the event that some but not necessarily all of the quantities in Table 4 may not be easily calculable, such as percentiles of running means, alternative quantities that are simpler to calculate may be useful. In their paper on the CAR model, Eerens and colleagues[18] gave some important insights from the work of Den Tonkelaar and van den Hout, who analysed measured concentrations in streets:

1. Whilst short-term concentrations are strongly affected by street geometry, the long-term annual concentration pattern is much less sensitive to the presence of buildings.
2. The ratio between the annual average values and the high percentiles of the frequency distribution of concentrations did not vary very much from street to street.

These points suggest that the use of surrogate statistics is proving to be a reasonably convenient way of deriving the high percentiles; by analogy we may expect a similar approach to the high percentiles of the running means. There is also the hope that such surrogates might be reasonably independent of particular towns or localities. They would then be useful in towns with no long-term monitoring records. However, Laxen[19] has suggested at a meeting of dispersion modellers that high percentiles may not have consistent relationships with the annual means. Clearly much more research in this area is indicated.

[17] RMS, *Atmospheric Dispersion Modelling: Guidelines on the Justification of Choice and Use of Models, and the Communication and Reporting of Results*, The Royal Meteorological Society/Department of the Environment, London, 1995.

[18] H.C. Eerens, C.J. Sliggers and K.D. Van Den Holt, *Atmos. Environ.*, 1993, **27B**, 389.

[19] D. Laxen, *Air Quality Standards in the UK National Strategy*, presented at the NSCA workshop, London, 26 November 1996, Techniques for Dispersion Modelling and Development of Good Practices, UK Dispersion Model Users Group Fourth Meeting, NSCA, Brighton, 1996.

### *Roll-back of Means*

There are occasions where the magnitude of emission reductions needs to be quickly estimated. The 'roll-back' equation can be used[16] to estimate a factor $R$ which is by how much emissions should be reduced to achieve a given reduction in the annual mean concentration (it assumes the background concentration is constant, *i.e.* independent of the source being evaluated):

$$R = \frac{\text{mean} - \text{standard}}{\text{mean} - \text{background}}$$

This approach may find application in assessing the control needed for sources influencing an air quality management area. It assumes that the standard is expressed as an annual mean. The equation means that a measure of the reduction can be obtained without dispersion modelling.

### *Reduced Percentiles*

Air quality data are very often better described by a log-normal distribution than by a normal one. Seinfeld[16] considers the fitting of log-normal distributions to air pollution data, and describes an extension of the above 'roll-back' idea to the controls needed to manage the high percentiles seen in the monitoring. Assuming that meteorological conditions are unchanged, background concentrations are negligible and all sources would be given the same amount of emission reduction, the standard geometric deviation of the distribution is unchanged when the emissions are varied, *i.e.* the slope of the line is unchanged by emissions control. Therefore reduction in emissions would move the plotted line of the log-normal distribution downwards to lower concentrations (plotted vertically, *cf.* page 691 of Seinfeld[16]) with the same slope. This translation on the graph is the same for all percentile concentrations; the expected shift in, say, a 98th percentile is the same as that expected for the 50th percentile (or median). Therefore as emission controls are applied, the lowering of the frequency at which a standard concentration is likely to be exceeded can be read from the graph (or calculated from the two parameters of the log-normal distribution that fits the data).

## 8 Role of Air Quality Modelling

Under the Environment Act 1995, as local authorities seek to review and manage air quality, there is likely to be increased use of air pollution modelling. The importance of modelling as a tool for air quality managers is manifest in European as well as DoE papers. Control of emissions in order to manage air quality relative to the standards is the *raison d'être* for dispersion modelling.

## European Dimension

An Explanatory Memorandum[20] explains that the EC Framework Directive on Air Quality Assessment and Management was a response to the 5th Action Programme. This programme saw a

> 'continuing need for legislative measures at Community level, particularly with respect to the establishment of fundamental levels of environmental care and protection; in addition it highlights the need for baseline data, statistics and indicators to assess environmental conditions and trends and where necessary to adjust and optimize policy. This Programme also advocates the setting of long-term objectives.'

These objectives sought to protect all people against recognized risks from air pollution and to establish permitted concentrations of air pollutants for environmental protection: this implied extension of the list of regulated substances, and monitoring and control of concentrations with regard to standards. Product standards and emission limits to be used in defining air quality objectives are in effect two of the tools that can be used to manage air quality; both operate through their control of what is released to the atmosphere.

The proposed Directive regards the parallel development of ambient air quality standards and emission/product standards as the best way of safeguarding air quality. It applies the principle of prevention via stringent air quality objectives sufficiently low that adverse effects are unlikely, based on present knowledge. Assessment of air quality will[20]

- 'allow identification of areas where specific actions need to be taken
- provide a direct way of measuring the impact of the measures implemented in order to reduce the emissions of atmospheric pollutants. Data on ambient air quality made available through this proposal can be linked with data on emissions by models. In the long run, the models can help to improve the emission inventories, air quality surveillance and the design of control measures. The detailed information collected through the assessment could also be used in relation to observed effects on environment and health.'

Two key points[20] were:

1. the need for means to measure the effectiveness of controls on emissions
2. the role of models

Models link measured air quality with emissions data. They can be used to improve emissions inventories and monitoring programmes, and assist in planning the control measures. The models will need to be capable of relating air quality to emissions. They must be deterministic, *i.e.* based upon physical principles, and be able to recognize individual sources and their influences.

[20] CEC, *Proposal for a Council Directive on Ambient Air Quality Assessment and Management*, (includes Explanatory Memorandum that accompanies the Proposed Directive), Commission of the European Communities, Brussels, COM (94) 109 final, 1994.

Points on modelling include:[20]

1. the need to set objectives to limit harmful concentrations
2. resources and methods to assess air quality should be balanced with the size of the problem
3. if modelling is used to assess air quality, attention to accuracy and criteria for use must be considered
4. mandatory measurements in the large conurbations and areas of poor/improving air quality, but measurements combined with modelling might be used where surveys suggest levels below 75% of limit values, whilst where values are below 50% of the limits, modelling or objective estimation may be used to evaluate the levels

## *DoE Discussion Paper*

The paper Improving Air Quality[21] set out the Government's view on early and effective action to improve air quality further:

> 'The Government believes that progress must be based on the further development of an effects-based approach, complemented by source-based controls. The essence of this approach is that the regulation and control regimes and strategies which aim to prevent specific emissions, by the introduction of cleaner technology . . . are unlikely to be sufficient on their own: it is important also to assess the impact of air pollution on human health and the environment, to set standards for specific pollutants in the light of this, and then to establish a comprehensive programme designed to secure those standards in the best possible way.'

An effects-based approach is focused on objectives, is cost effective and allows environmental gains from regulatory controls to be quantified and judged:[21]

> 'The essential requirements for an effects-based approach are an ability to set standards for ambient air quality which define satisfactory air quality and identify the impact of various levels of a contaminant; and secondly, a framework of policies and procedures which allows air quality to be managed to secure those standards.'

In some urban areas, development can put air quality at risk, and the paper cites the East Thames Corridor which was subject to an air quality management study by HMIP. There was concern that industrial development, with significant new sources, would lead to breaches of the EC Air Quality Directive limit value for $NO_2$. The study's conclusion was that it was not likely that the limit value would be breached, but it emphasized that

[21] DoE, *Improving Air Quality*, a discussion paper on Air Quality Standards and Management, Department of the Environment, London, 1995.

> 'An important issue in areas of urban and industrial development therefore is the need in the planning framework to consider all sources of pollutant emissions and to assess their respective future development in an integrated way.'

This study had to identify the additional contribution due to a proposed development, and modelling was the technique employed. This role of modelling reminds us of the European Objectives discussed earlier.

Despite existing legislation, upgrades in process authorizations by HMIP and new technology on vehicles,

> 'Certain areas may still experience episodes of high pollution. Many factors will play a part, including weather conditions and local topography, but the critical factor is the overall sum of emissions from different sources and their interaction, with each managed in a different framework. Urban areas, where the concentration of sources is greatest, are most likely to be at risk. Even small 'hot-spots' may therefore affect many people.'

Even when national policies are delivering reductions in overall emissions, local developments may congregate sources together and could worsen air quality in some areas. Local air quality management therefore received significant attention in the Environment Act 1995. Reviews of air quality will need to be forward looking and allow for planned developments.

The paper[21] considers the matter of fairness:

> 'The impact of air quality controls in an area may not be even-handed.... The contribution which a sector makes should take into account the costs and benefits of measures to reduce emissions from other relevant sectors.'

The paper[21] seeks to establish means of managing air quality that recognize the multitude of different sources within an urban area, and are both cost effective and even-handed. Modelling can recognize individual source contributions. It can recognize their contributions aggregated into sector type, *e.g.* traffic, industrial or domestic sources, as in the East Thames Corridor study cited earlier.

Smokeless zones regulated domestic emissions of smoke and sulfur dioxide, and the BATNEEC procedure was used for industrial controls. However, in 1994, before the advent of local air quality management under the 1995 Act, there was no

> 'ready means of balancing contributions from industrial sources with those from domestic and mobile sources.'

In the interests of fairness, there should be a regulatory means

> 'to ensure any necessary overall reduction in emissions is even-handedly distributed. In practice, any action should, of course, be targeted and proportionate.'

Concerning traffic management and planning

> 'Closer attention may need to be paid to air quality in areas which are at risk of episodes of high air pollution. One possibility would be to ask local authorities in such areas to assess the likely effects of the development plan against a model of air pollution in the area.'

Local air quality reviews and assessments are new duties for local authorities under the Environment Act 1995. Three basic tools that local authorities will be able to use in meeting their statutory duties are dispersion modelling, monitoring and emission inventories. The Department would not promote a single national model, but it did anticipate providing advice on the suitability of models for various purposes.[21] Advice on all three tools is now available or is nearing publication.

Air quality management studies using these tools include emissions and modelling in Belfast to help plan compliance with the smoke/$SO_2$ Directive, a comparison of power station and urban contributions to $SO_2$ in Lincoln after concern at possible damage to the Cathedral, a study of emissions in London and the East Thames Corridor Study cited earlier.[21] As a result of the last study, additional controls were imposed on a major emitter in the area.

## *Strategic Policies*

With this background, in January 1995 the Government published a paper on its strategic policies for air quality management.[22] The paper has much in common with the points discussed above. These plans for local air quality management, the reviews and assessments, the need for guidance and the designation of air quality management areas has since become law. In addition to the Environment Act 1995, another important document is the Strategy,[1] which contains much technical information on the priority pollutants and their sources of emission. It also lists the standards and objectives for air quality.

With regard to monitoring and public information, the paper[22] includes the statements:

- 'In managing air quality at a local level, local authorities may need access to modelling systems which link monitoring with inventories of source emissions. The Department will provide assistance with the development of local emissions inventories in areas throughout the UK and will also provide access to models for use in association with such inventories.
- 'The Government believes that it would be appropriate to extend its monitoring and modelling activities to provide more comprehensive assessments of population exposure.'

The DoE has funded a significant effort by local authorities to test and evaluate a number of models; reports from these trials will assist the drafting of advice to

[22] DoE, *Air Quality: Meeting the Challenge, The Government's Strategic Policies for Air Quality Management*, Department of the Environment, London, 1995.

local authorities. Such advice will describe likely best practice in this area.

### *Forecasts of Episodes*

In areas of poor air quality[20]

> 'Member States shall make short term plans for actions to be taken in cases when an exceedance is forecast in order to reduce the likelihood of exceedance and to limit its duration.'

With regard to public information and episodes, the paper[22] mentions that air quality forecasts for the following 24 hours are available; the Government aims to make air quality information as accessible as possible to the public: monitoring data are now available on the Internet.

Forecasting air quality has been alluded to already in a European context:[21]

> 'Once standards have been attained, it remains likely that occasional episodes of high pollution, or the risk of them, will from time to time arise. A response mechanism, sensitive to local circumstances, should therefore be available for use when required to tackle such problems.'

We draw these discussions to a close with two pertinent extracts from the Environment Act (1995), which says that local authorities have duties to review air quality and inform the public:

- Reviews '81. (1) Every local authority shall from time to time cause a review to be conducted of the quality for the time being, and the likely future quality within the relevant period, of air within the authority's area.'
- Communication '82. (2). . . regulations. . . may make provision—. . . (k) for or connection with the communication to the public of information relating to quality for the time being, or likely information relating to quality for the time being, or likely future quality of the air.'

## 9 Conclusions

1. Improvements in air quality depend upon public agreement on the targets to be attained and the reasons for seeking to achieve them. The UK National Air Quality Strategy[1] represents a culmination of many research papers and reviews. It sets out the likely trends in each main pollutant in the next decade, reviews their sources, environmental and health effects, and describes standards and objectives. It heralds a new approach to air quality management based upon an effects-based approach.

2. The Environment Act 1995 has put into law many of the ideas on local air quality management seen in earlier papers. The reviews and assessments of local air quality, and the provision to prepare action plans in pollution 'hot-spots' known as air quality management areas, provide a new basis for local authority control of air quality.

3. The requirement from dispersion modelling lies in the need for cost effective and equitable management of the various emission sectors. Modelling is an indispensable tool because it provides a way of identifying the contributions from individual pollution sources. Dispersion modelling, when combined with the latest developments in emission inventories, is a planning tool, because it allows alternative policies to be compared and assessed in an objective way.

4. The Royal Meteorological Society has offered guidelines on the choice and use of models in atmospheric dispersion. They highlight the importance of the audit trail.

5. As an example of how one local authority has approached air quality management, the Kent study[23] contains dispersion modelling, and the models used are outlined in the appendix to their report. The approach used enables individual sources to be modelled.

6. Emission inventories, monitoring and dispersion modelling are the three techniques upon which air quality reviews, assessments and action plans will rely. It is important that these are on the public record, and that the work has a clear audit trail, so that proposals and decisions can be explained and supported with sound evidence. Plans to improve air quality in local authority areas can then be agreed.

[23] Kent, *The Kent Air Quality Management System Final Report*, Kent Air Quality Partnership, Kent County Council, Maidstone, Kent, 1996.

# Emission Inventories

DAVID HUTCHINSON

## 1 Introduction

Smoke and sulfur dioxide pollution from domestic coal burning, industry and power stations was the cause of the great London smog in December 1952. Now urban air quality is again causing public concern, largely as a result of emissions from motor vehicles. An atmospheric emissions inventory is a schedule of the sources of pollution within a particular geographical area. The inventory usually includes information on the amount of the pollutant released from major industrial sources, and average figures for the emissions from smaller sources throughout the area. Emission inventories are an essential tool in the management of local air quality. Whilst monitoring, such as that carried out through the UK's automatic air quality monitoring network,[1] shows the extent of air pollution, emission inventories identify the sources and help in the development of abatement strategies.

The preparation of urban emission inventories is not a new idea in the UK. Two of the earliest urban inventories were prepared for Sheffield[2] and Reading[3] in the mid-1960s. The first London inventory was prepared by the Scientific Branch of the former Greater London Council by Ball and Radcliffe in 1979.[4] This was an inventory of sulfur dioxide ($SO_2$) emissions.

Although sulfur dioxide levels had been falling in London since the 1950s, there was concern because concentrations were still high compared to other cities. There was a danger that London would not comply with the then proposed European Community Directive on air quality, although the numerical values of the proposed standards were still under discussion at the time. There was also concern about the adverse effect of $SO_2$ on health and vegetation, and the damage caused to the fabric of buildings.

Ball and Radcliffe identified five main applications for an urban emissions inventory, which are as valid today as they were in 1979. They are:

[1] J. S. Bower, G. F. J. Broughton, P. G. Willis and H. Clark, *Air Pollution in the UK: 1993/94*, AEA Technology: National Environmental Technology Centre, Culham, 1995.

[2] A. Garnett, *Trans. Inst. Br. Geog.*, 1967, **42**, 21.

[3] K. J. Marsh and M. D. Foster, *Atmos. Environ.*, 1967, **1**, 527.

[4] D. J. Ball and S. W. Radcliffe, *An Inventory of Sulfur Dioxide Emissions to London's Air*, Research Report 23, Greater London Council, London, 1969.

- An emissions map, showing the geographic distribution of emissions, can be an important aid in land use planning by identifying parts of the region that are likely to be subject to high levels of pollution, and the location of pollution sources in relation to sensitive areas.
- An emissions inventory can help in estimating the cost of introducing controls, and identifying who should bear those costs.
- The design of monitoring networks is important, if meaningful data are to be obtained. The data may be required to assess the exposure of the population to a particular pollutant, or to demonstrate compliance (or non-compliance) with air quality standards. An emissions inventory will indicate, for example, where the highest concentrations of pollution are likely to be found, or which areas are the most representative. It can thus help to ensure that monitoring equipment is appropriately located.
- Ideally, it should be possible to use an emissions inventory, in conjunction with an atmospheric dispersion model for the area, to predict short-term pollutant concentrations at ground level during forecast adverse weather conditions. When these techniques are fully developed, they could be used to alert authorities to possible air pollution incidents and to determine strategies for avoiding them.
- An emissions inventory can be used, alone or in conjunction with an atmospheric dispersion model, to assess trends in air quality. By altering the input to the emissions inventory in a way that simulates future conditions (such as a change in fuel use), it is possible to make predictions about the impact on air quality.

In addition, air quality monitoring can be used to validate emission inventories, indicating where emission factors (see below) may be too high or too low.

In the United States, regulations requiring the annual reporting of emissions have been in place since the late 1970s. Under Title 40, Code of Federal Regulations, Part 51.321, states are required to prepare and submit annual reports to the US Environmental Protection Agency (US EPA) on emissions of $SO_2$, carbon monoxide (CO), oxides of nitrogen ($NO_x$), volatile organic compounds (VOC), particulate matter and lead within their boundaries. The Clean Air Amendment Act 1990 requires the development of '. . . comprehensive, accurate and current. . .' inventories from all sources. The US EPA's Office of Air Quality Planning and Standards develops and maintains emission estimating tools to support the preparation of emission inventories by federal, state and local agencies, consultants and industry.[5] Emissions inventories are also being prepared in many other parts of the world, from Mexico[6] to Korea.[7]

[5] S. Sasnett and D. Misenheimer, *Regulatory Requirements for Emissions Reporting to EPA*, proceedings of the conference on The Emissions Inventory: Programs & Progress, Research Triangle Park, NC, 11–13 October 1995, Air & Waste Management Association, Pittsburgh, 1996, pp. 65–70.

[6] R.J. Dickinson, L.J. Markovich, M.E. Wolf, W.R. Oliver and L.W. Creelman, *Development of a National Inventory Program for Mexico*, proceedings of the conference on The Emissions Inventory: Programs & Progress, Research Triangle Park, NC, 11–13 October 1995, Air & Waste Management Association, Pittsburgh, 1996, pp. 687–701.

[7] Y.-J. Kim, *Preparation of Emissions Inventories and Establishment of the National Emission*

## The European Context

There are a substantial number of international agreements which require the preparation of emission inventories. The Convention on Long Range Transboundary Air Pollutants was adopted in Geneva in 1979 within the framework of the United Nations Economic Commission for Europe (UNECE). States are required to report emissions data to the Executive Body of the Convention in order to meet their obligations to comply with the Convention Protocols and to assess which countries need to adopt the most stringent abatement strategies. The Protocols are:

- Protocol on Long-term Financing of the Co-operative Programme for Monitoring and Evaluation of the Long Range Transmission of Air Pollutants in Europe
- Protocol on the Reduction of Sulfur Emissions or their Transboundary Fluxes (1985)
- Second Protocol on Further Reduction of Sulfur Emissions (1994)
- Protocol concerning the Control of Emissions of Nitrogen Oxides or their Transboundary Fluxes (1988)
- Protocol concerning the Control of Emissions of Volatile Organic Compounds or their Transboundary Fluxes (1991)

The parties to the Convention, including the UK, are required to submit details of their annual national emissions of $SO_2$, $NO_x$, methane ($CH_4$), non-methane volatile organic compounds (NMVOC), ammonia ($NH_3$) and CO.

The Co-operative Programme for Monitoring and Evaluation of the Long Range Transmission of Air Pollutants in Europe (EMEP), established under the Convention, in turn set up a Task Force on Atmospheric Emissions Inventories in 1991 with a variety of objectives including the preparation of an Atmospheric Emissions Inventory Guidebook.[8] The aim of this was to increase the reporting of emissions by the signatories to the Convention. It also established a 50 kilometre × 50 kilometre grid, known as the EMEP grid, for reporting and analysis purposes, and 11 main source categories. These are:

1. Public power, cogeneration and district heating plants
2. Commercial, institutional and residential combustion plants
3. Industrial combustion
4. Production processes
5. Extraction and distribution of fossil fuels
6. Solvent use
7. Road transport

*Inventory System of Air Pollutants in Korea*, proceedings of the conference on The Emissions Inventory: Programs & Progress, Research Triangle Park, NC, 11–13 October 1995, Air & Waste Management Association, Pittsburgh, 1996, pp. 683–686.

[8] *Atmospheric Emissions Inventory Guidebook*, ed. G. McInnes, European Environment Agency, Copenhagen, 1996.

8. Other mobile sources and machinery
9. Waste treatment and disposal
10. Agriculture
11. Nature

In 1985 the European Community set up a programme for the collection of information on the state of the environment and natural resources in the Community. This programme was given the name CORINE—CO-oRdination d'INformation Environmentale—and included a project to gather and organize information on emissions to the atmosphere which were relevant to acid deposition. This is known as CORINAIR. The programme included the preparation of an inventory as well as the preparation of a default emission factor handbook (to be used in the absence of more specific data) computer software and a standard set of descriptions of both sources and activities. This is known as the Selected Nomenclature for Air Pollution (SNAP). The first CORINAIR inventory was completed in 1990.

The Framework Convention on Climate Change also requires that national inventories of anthropogenic emissions of greenhouse gases are to be prepared. The Inter-governmental Panel on Climate Change (IPCC), with the support of OECD and the International Energy Agency, has prepared Guidelines for National Greenhouse Gas Inventories.[9]

The European Environment Agency was set up in 1993 in order to provide the Member States of the European Community 'with objective, reliable and comparable information at the European level, enabling them to take the requisite measures to protect the environment, to assess the results of such measures and to ensure that the public is properly informed about the state of the environment'. As part of its first work programme the Agency designated five European Topic Centres to address the problems of inland waters, the marine and coastal environment, nature conservation, air quality and air emissions. The work programme for the European Topic Centre on Air Emissions (ETC/AEM), in turn, requires the development of emissions inventory guidelines at various levels, the compilation of a European emissions inventory and a review of the CORINAIR methodology. The Agency also took over the editing and publication of the EMEP Atmospheric Emission Inventory Guidebook referred to above.

## *The United Kingdom Context*

As part of the government's continuing programme of air pollution studies, the Department of the Environment (DoE) has developed a national inventory of air pollution sources and the type and quantity of the pollutants they emit. This is called the National Atmospheric Emissions Inventory (NAEI). The coverage of this inventory has been expanded as emissions of pollutants have grown or as evidence of their adverse effects has accumulated. Initially, only emissions of black smoke and $SO_2$ were estimated but now 12 pollutants or pollutant groups are covered. They are $CO_2$, $SO_2$, black smoke, CO, $NO_x$, $CH_4$, NMVOC, $NH_3$,

[9] *IPCC Guidelines for National Greenhouse Gas Inventories*, OECD, Paris, 1995.

**Table 1** Estimated emissions in the United Kingdom 1994 (percentages)

| *UNECE source category* | *Sulfur dioxide* | *Nitrogen oxides* | *Carbon monoxide* | *Carbon dioxide (as carbon)* | *Methane* | *Volatile organic compounds* | *Black smoke* |
|---|---|---|---|---|---|---|---|
| 1 Public power, cogeneration and district heating | 65 | 24 | 0 | 30 | 0 | 0 | 4 |
| 2 Commercial, institutional and residential combustion | 6 | 5 | 6 | 21 | 1 | 2 | 23 |
| 3 Industrial combustion | 24 | 10 | 1 | 26 | 0 | 1 | 4 |
| 4 Non-combustion industrial processes | 0 | 0 | 0 | 0 | 0 | 19 | 0 |
| 5 Extraction and distribution of fuels | 0 | 5 | 1 | 1 | 21 | 11 | 0 |
| 6 Solvent use | 0 | 0 | 0 | 0 | 0 | 31 | 0 |
| 7 Road transport | 2 | 49 | 89 | 21 | 1 | 29 | 58 |
| 8 Other mobile sources | 2 | 7 | 1 | 2 | 0 | 1 | 1 |
| 9 Waste treatment and disposal | 0 | 0 | 1 | 0 | 48 | 1 | 10 |
| 10 Agriculture | 0 | 0 | 0 | 1 | 29 | 0 | 0 |
| 11 Forests | 0 | 0 | 0 | 0 | 0 | 4 | 0 |
| TOTAL (thousand tonnes) | 2718 | 4218 | 4833 | 149[a] | 3876 | 2117 | 426 |

Source: Digest of Environmental Statistics: 1996, No. 18.
[a]Carbon dioxide in millions of tonnes.

nitrous oxide ($N_2O$), lead, other heavy metals and halogens. In addition, HFCs and particulate matter less than 10 $\mu$m aerodynamic diameter ($PM_{10}$) are being considered for inclusion in the inventory. The NAEI is maintained for the DoE by the National Environmental Technology Centre and the results are published in the DoE's annual Digest of Environmental Statistics[10] and are summarized in Table 1.

The purpose of the National Atmospheric Emissions Inventory is:

- as an input to discussions with various international bodies, including those mentioned above
- as an input to UK policy-making with respect to pollution abatement and control
- to assist in judging the effectiveness of existing policies
- as an aid to the interpretation of air quality measurements
- as an input to atmospheric dispersion models
- for general public information

The relationship between the UK National Atmospheric Emissions Inventory and the urban inventories now being prepared is illustrated in Figure 1.

Smoke and sulfur dioxide pollution from domestic coal burning, industry and power stations was the cause of the great London smog in December 1952. Now urban air quality is again causing public concern, largely as a result of emissions from motor vehicles. Although the present situation is less serious than that in the 1950s, results from the automatic air quality monitoring network show that concentrations of pollutants exceed non-mandatory health guidelines in many British cities.

The government is developing a new strategic framework for air quality management. Its approach is set out in the consultation draft of 'The United Kingdom National Air Quality Strategy', published in 1996.[11] This contains key proposals relating to the development of air quality standards and objectives for nine pollutants, as well as the creation of air quality management areas. The legal framework is provided by Part IV of the Environment Act 1995 which deals with air quality.

The starting point for local authorities is the review of local air quality and its assessment against standards and objectives. The results of this appraisal will determine what, if any, further action may be required, such as the establishment of an Air Quality Management Area. The DoE is currently drawing up guidance on review and assessment for local authorities. Quality assessments will probably make use of these three main tools: air quality monitoring, local emission inventories and numerical modelling. The nature and complexity of these assessments will vary, of course, depending on factors such as local pollution sources, topography and the risk of standards being exceeded.

The DoE's Air and Environment Quality Division (DoE AEQ) funded the preparation of an emission inventory for the West Midlands metropolitan area,

[10] Department of the Environment, *Digest of Environmental Statistics: No. 18, 1996*, HMSO, London, 1996.

[11] Department of the Environment, *The United Kingdom National Air Quality Strategy—Consultation Draft*, Department of the Environment, London, 1996,

**Figure 1** The relationship between the National Atmospheric Emissions Inventory and urban missions inventories

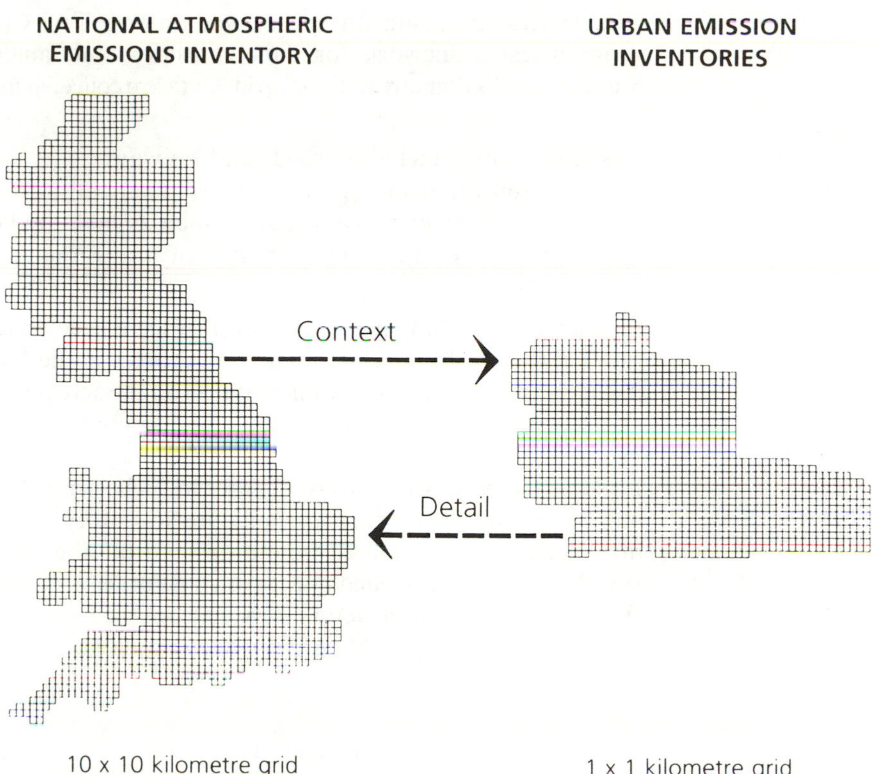


which was published in November 1996.[12] One of the objectives was to develop methodologies which could be applied in other cities, and work has started on the preparation of inventories in Bristol, Portsmouth/Southampton, Liverpool/Manchester, Glasgow and Swansea/Neath/Port Talbot as part of a series of ten further inventories being prepared over the next two years (the Ten Cities Programme). They are also being prepared by the London Research Centre on behalf of DoE AEQ, with the active support of the constituent local authorities.

Space does not allow a full discussion of the methodologies used in preparing Europe-wide inventories, the UK National Atmospheric Emissions Inventory and also the urban emission inventories. The next section has therefore been confined to a discussion of urban atmospheric emissions inventory procedures.

## 2 Urban Emission Inventory Procedures

The London sulfur dioxide inventory[13] and subsequent London studies[14,15] established the general methodology used in preparing the West Midlands

[12] D. Hutchinson and L. Clewley, *West Midlands Atmospheric Emissions Inventory*, London Research Centre, London, 1996,

[13] D. J. Ball and S. W. Radcliffe, in ref. 4, p. 5.

[14] D. J. Ball, C. Fernandes, D. Hutchinson, H. Kostanowicz, A. Onslow and C. E. Wright, *Energy Use in London*, Greater London Council, London, 1981.

[15] M. Chell and D. Hutchinson, *London Energy Study*, London Research Centre, London, 1993.

Atmospheric Emission Inventory and in the Ten Cities Programme. The geographical framework for data collection and analysis is the Ordnance Survey's 1 × 1 kilometre national grid. Data are collected for three types of sources:

- Line sources including roads and railways
- Area sources including emissions from agricultural and other land, and low intensity emissions from sources such as building heating systems
- Point sources including high intensity emissions from industrial plants

The number of emission sources which have been continuously monitored is small in relation to the total number of sources in the UK as a whole or in any particular urban area. Hence information on the actual emissions occurring from an individual source is only available in a limited number of cases in any area. Clearly it is impossible to start measuring every emission source in an area like London, the West Midlands or Greater Manchester with a population of several million in order to prepare an emissions inventory. The majority of emissions must therefore be estimated from other information such as fuel consumption, vehicle kilometres travelled (VKT) or some other measure of activity relating to the emissions. Emission factors, derived from the results of the monitoring which has been undertaken, are then applied to the activity data in order to estimate the likely emissions:

$$\text{Activity rate} \times \text{Emission factor} = \text{Emission}$$

Figure 2 illustrates the general procedure followed in order to convert information on activities, such as road traffic, into estimates of emissions to the atmosphere.

For many of the pollutants of concern, the major source of emissions is the combustion of fossil fuels. Consequently the collection and analysis of fuel consumption statistics plays an important part in the preparation of emission inventories. However, it is important to consider the differences between consumption and fuel deliveries when making use of the available data. Most of the readily available statistics relate to fuel deliveries which, in many cases, relate closely to consumption. However, in the case of fuels which may be stockpiled, such as coal, there may be significant differences between delivery and consumption. In the case of transport fuels, there may be significant geographic differences between the point of delivery and where the fuel is used. The most striking example is London's Heathrow Airport, to which some 3.7 billion litres of aviation fuel is delivered annually. Only a small fraction of this fuel is used within the London area whilst the remainder is used on journeys to the far corners of the Earth.

The pollutants and pollutant groups covered in the current UK urban inventories include:

- Sulfur dioxide ($SO_2$)
- Oxides of nitrogen ($NO_x$)

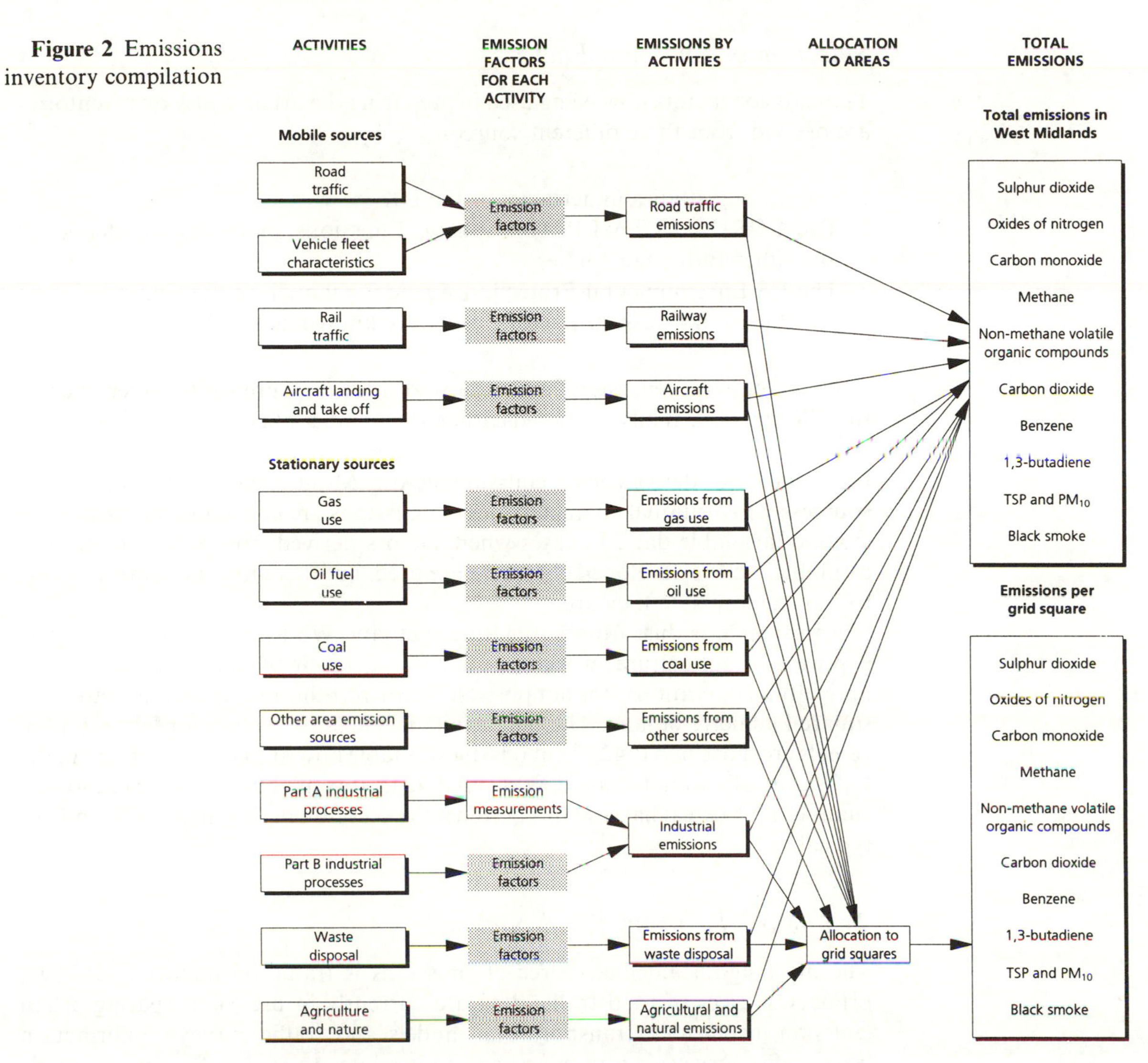


**Figure 2** Emissions inventory compilation

- Carbon monoxide (CO)
- Methane ($CH_4$)
- Non-methane volatile organic compounds (NMVOC)
- Carbon dioxide ($CO_2$)
- Benzene
- Buta-1,3-diene
- Total suspended particulates (TSP)
- Particulate matter less than 10 mm aerodynamic diameter ($PM_{10}$)
- Black smoke

Black smoke is included for the sake of continuity with earlier studies. A national smoke and $SO_2$ monitoring network has operated since the 1960s.

## Selection of Emission Factors

The emission factors now being used in preparing the urban emission inventories are derived from three different sources:

1. Monitoring studies undertaken in the UK
2. The CORINAIR/EMEP Atmospheric Emissions Inventory Guidebook[16] and other European studies
3. The US Environmental Protection Agency's manual on the Compilation of Air Pollution Emission Factors, generally known as AP-42[17]

Where possible, emission factors derived from measurements undertaken in the UK are used. Whilst, as noted earlier, the number of emission sources which have been continuously monitored is limited, in many cases enough sources have been studied to derive robust emission factors. Monitoring of further emission sources and additional characteristics of emissions are constantly adding to the stock of available data. In cases where factors derived from UK data are not available, factors developed elsewhere are used, but there may be some doubt as to how appropriate they are.

An example is dust from agricultural activity. Whilst there are data on the exposure of agricultural workers to dust as a result of harvesting and other agricultural operations, it is not possible to convert this into emission factors per square kilometre of land. The only available emission factors for harvesting of cereals are those in AP-42,[18] but it is questionable how applicable these are in the UK's generally wetter climate. Whilst AP-42 factors might be appropriate for the summer 1995 cereal harvest, this period was exceptionally dry in the UK and not typical.

## Traffic on the Main Road Network

The most significant line source of emissions is traffic on major roads. Two primary sources of road traffic data are currently in use for preparing urban emission inventories: transportation models and traffic surveys. Information from traffic surveys is attractive because it relates to *real* traffic on *real* roads whereas transportation models are a computerized reflection of the real conditions. However, traffic surveys generally have the distinct disadvantage that they do not cover the required study area completely and/or in enough detail, usually having been undertaken for a totally unrelated purpose. On the other hand, transportation models are available for most of the main urban areas and are comprehensive.

The models provide estimates for morning peak traffic flows, and often for other periods of the day, on a network of major roads throughout London, the

[16] Ref. 8, Part B, Methodology chapters.

[17] *Compilation of Air Pollution Emission Factors*, AP-42 5th edn., United States Environmental Protection Agency, Office of Air Quality Planning and Standards, Research Triangle Park, NC, USA, 1995.

[18] Ref. 17, Section 9.3.2.

West Midlands, the Manchester conurbation and other areas. The network is divided into links and for each link there is a traffic flow, a link length, the traffic speed or the travel time for the link which can be used to calculate speed, and the national grid coordinates of the link. The coordinates can be used to allocate the links to the appropriate 1 × 1 kilometre grid squares.

In order to estimate the annual fuel use and emissions along each link and within each grid square, it is necessary to know the proportions of any links within a particular grid square, the link lengths, the annual flows, speed and the composition of the traffic (heavy and light commercial vehicles, buses, cars/taxis and motorcycles as well as the split between diesel and petrol engined vehicles). The models do not generally contain all the information required, for example on traffic composition. Where this is not available, the results of traffic surveys can be used to ascribe an average traffic composition to each link in the road network. The proportions of the vehicle fleet within each of these categories which are petrol and diesel engined are estimated from the vehicle licensing records. The total annual VKT for each category of vehicle can then be combined with emission factors in order to calculate the total emission for each link. These can then be aggregated for each grid square. Figure 6 below, which shows the emissions of nitrogen oxides from traffic in the West Midlands, has been compiled in this way.

## *Vehicle Emission Factors*

The fuel consumption and pollutant emissions of vehicles depend on a great variety of factors including vehicle and engine type, vehicle age and tuning condition, climatic conditions and vehicle temperature. The actual conditions of vehicle operation resulting from traffic conditions (congested urban roads, free-flowing rural roads, *etc.*), the vehicle use (short or long trips, frequency of trips, *etc.*) and the driver's behaviour (gentle or aggressive driving) give rise to some of the widest variations in consumption and emissions. Information on the effects of vehicle operating conditions and of use on emissions is much more limited than one would wish.

As the transport models currently being used in preparing inventories provide data on vehicle speeds, it was highly desirable to find emission factors that could make use of this rather than generalized 'urban cycle' factors. A large scale experimental study was conducted between 1989 and 1992 under the European Community's DRIVE programme to characterize the actual conditions of car operation and to improve knowledge of fuel consumption and exhaust emissions.[19] The study was conducted jointly by the Transport Research Laboratory (TRL) in the UK, TÜV Rheinland in Germany and the Institut National de Recherche sur les Transports et leur Sécurité (INRETS) in France. The first step was to undertake a substantial survey of the operating characteristics of cars in urban areas in order for INRETS to develop 14 typical urban driving cycles. The cities studied were:

[19] P. Jost, D. Hassel, F.J. Weber and K.S. Sonnborn, *Emission and Fuel Consumption Modelling Based on Continuous Measurement*, Deliverable Nr. 7 of the DRIVE PROJECT V 1053, Commission of the European Communities, Brussels, 1992.

- London and Derby in the UK
- Cologne and Krefeld in Germany
- Grenoble and Marseilles in France

The driving cycles were then reproduced on chassis dynamometers in each of the participating laboratories and fleets of vehicles tested in order to measure the fuel consumption and emissions. The vehicles were selected from three age bands which reflected the introduction of progressively tighter emission control requirements for petrol engined cars, and three different engine sizes, as well as diesel engined cars. The vehicles were tested in their delivered state, *i.e.* without any prior maintenance or repair. The results of this study were found to provide the type of vehicle speed-related data required to convert the transport model output into emissions of NMVOC, CO, $CO_2$ and $NO_x$. Figure 3 shows the relationship between car speeds and emissions of $NO_x$ and CO. Unfortunately, the same type of information is not available for heavy-duty trucks, and generalized urban and other driving cycle emission factors have to be used.

### *Other Vehicle Emissions*

In addition to the traffic on the main road network, there is a substantial volume of traffic on local roads. In the West Midlands, the West Midlands Joint Data Team, which manages the transportation model, was able to provide an estimate of the residual minor road traffic. Emissions from minor roads are treated as an area rather than a line source. The emissions inventories also include estimates of emissions from cold start, hot soak, diurnal emissions from vehicle fuel systems, refuelling, tyre and brake wear, and road dust.

### *Verification of Road Transport Fuel Use*

Wherever possible, estimates are verified using alternative data sources. For example, the estimates of total vehicle fuel consumption were compared with data on vehicle fuel deliveries in the West Midlands. Information on petroleum fuel deliveries, including petrol and diesel fuel, is compiled by the Petroleum Industries Association (PIA). In the West Midlands study, the PIA petrol sales figure and the calculated petrol use were within 2%. A similar cross-check carried out as part of the earlier London Energy Study found the petrol sales figure and the calculated petrol use to be within 1%.[20] The close correlation between the figures in each case has been taken as generally confirming the validity of the methodology.

### *Emissions from Other Transport Modes*

As with road transport, an estimate of the emissions is required for each link in the rail network and for each kilometre grid square. The use, by passenger services, of each link by different types of locomotives and different lengths of

[20] D. Hutchinson and L. Clewley, Ref. 12, p. 19.

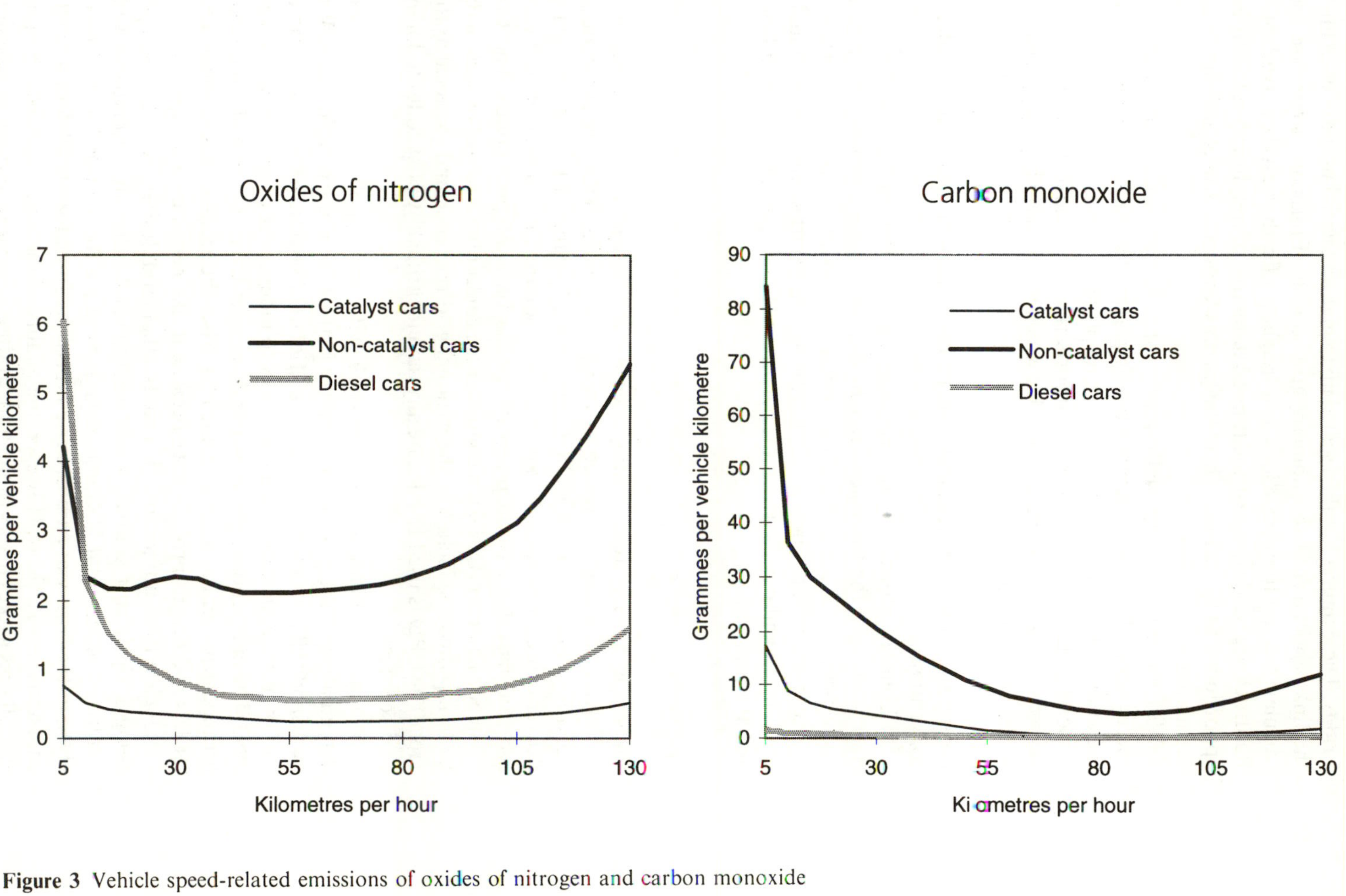


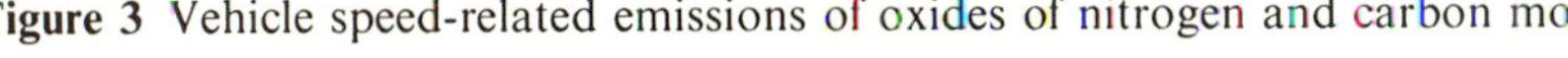

**Figure 3** Vehicle speed-related emissions of oxides of nitrogen and carbon monoxide

train is quantified, both during a peak hour and a non-peak hour Monday–Friday. These figures are then scaled up to give an annual usage of each link, and the links apportioned to kilometre grid squares. Emission factors are used to calculate the resultant emissions. For freight traffic, the best data available from Railtrack are used to assess the tonne-kilometres of freight moved. This is then apportioned to those lines where freight was known to travel and the emissions calculated.

With air travel, it is especially important to recognize that the concern is with energy use and emissions actually taking place within the area of study, not with the total energy use and emissions taking place as a result of the existence of the airport. In London there are six airports ranging from Heathrow, the world's largest international airport, to Kenley which is only used by gliders. In the West Midlands there are airports at Birmingham and Coventry. The emissions occurring with 1000 metres of the ground as a result of the aircraft landing, taxiing and taking-off are calculated in accordance with EMEP and CORINAIR conventions.

In London, information was gathered from the Port of London Authority on river traffic and the tonnage of freight shipped on the Thames. From these, an assessment was made of the contribution to atmospheric pollution of transport by water in London. There is very little water transport in the West Midlands, but it will be a significant factor in the port cities of Liverpool, Bristol, Southampton and Portsmouth.

## *Area Emission Sources*

Emissions arising from heating systems in homes and in commercial and industrial buildings are treated as area sources unless they are substantial and occur at a specific location. Data are compiled by the PIA of all petroleum fuel deliveries occurring in each 1 kilometre grid square. Fuel sales by British Gas and the coal industry can be provided on the basis of post codes. The Central Postcode Director (CPD) is a computer file containing a record of each of the 1.7 million postcodes, including the national grid reference and the local authority area. The CPD is used to allocate any data identified by post codes to kilometre grid squares.

A major problem is now occurring with gas supply data as a result of deregulation of the gas supply industry in the UK. For example, in the Walsall and Sandwell district of British Gas's Midlands Region, 63% of the commercial and industrial gas market had been lost to independent suppliers by November 1993. The independent gas suppliers do not yet have any central data collection arrangements similar to those of the PIA. Also, the industry has generally become far less inclined to provide data because it has become so competitive.

It is clear from numerous other studies that landfill sites are widely accepted as being a significant source of methane. Several different estimates have been published of the total UK emissions of methane from landfill sites. The figure given in the National Atmospheric Emissions Inventory is 48% of all methane emissions (see Table 1). However, the quality of the data available about many landfill sites makes it very difficult to produce accurate estimates of emissions on a grid basis. Whilst the locations of all sites are, or should be, known, records of the start and completion dates, the nature of the material filled and other data are incomplete. In the past, private operators did not always keep records and, even though landfill regulations have been tightened considerably, it is often impossible to reconstruct the history of individual sites.

The Institute of Terrestrial Ecology's Land Cover Map provides information on 25 types of land cover including agricultural land, woodland and water areas, as well as urban and suburban land, on a 25 metre grid.[21] This is derived from Landsat Thematic Mapper data. Whilst the data are not required on the 25 metre grid as such for the present inventories, it does allow the percentage of each land cover type occurring in each 1 kilometre grid square to be calculated. This information can then be used in conjunction with emission factors to calculate the emissions from non-urban land uses.

The Census of Agriculture, which is undertaken every June, provides information on crops cultivated and livestock.[22] There are, for example, 13 727 beef or dairy cattle and calves in the West Midlands as well as sheep, pigs and 176 000 poultry, even though this is a predominantly urban area. Emission factors are used to estimate both the enteric emissions of methane and the emissions from animal waste. There are 15 600 hectares of cereals being grown in the area. However, as already noted, there are no UK derived emission factors for the dust which results from harvesting and other agricultural operations.

## *Point Source Emissions*

Many of the emissions to the atmosphere resulting from industrial processes and the combustion of fossil fuels are not uniformly spread across urban areas but concentrated at particular points. These point sources include central heating plants serving large groups of buildings, such as hospitals, and boiler plants supplying process heat to industry. They also include industrial processes which require authorization under the Environmental Protection Act 1990.

This Act introduced a single system of control over 'prescribed processes' in industry and commerce which are likely to result in the release of pollutants to the environment (air, water and land). A common system of authorization, enforcement and public access to information applies to the Environment Agency and local authorities. The Environment Agency is responsible for the control of about 2200 processes that are most likely to cause serious pollution, which are referred to as 'Part A processes', and local authorities control more than 12 000 less threatening processes, referred to as 'Part B processes'. The Environmental Protection (Prescribed Processes and Substances) Regulations 1991, lists both the processes and substances coming under Environment Agency control and those under local authority control.

The Environmental Protection Act 1990 requires operators of Part A processes to obtain specific authorization, to monitor the release of pollutants

[21] R. M. Fuller, G. B. Groom and A. R. Jones, *Photogram. Eng. Remote Sensing*, 1994, **60**, 553.

[22] Ministry of Agriculture, Fisheries and Food, The Scottish Office Agriculture and Fisheries Department, Department of Agriculture for Northern Ireland and Welsh Office, *The Digest of Agricultural Census Statistics—United Kingdom 1995*, HMSO, London, 1996.

and to submit information in order to demonstrate compliance with the standards set by the Environment Agency. Failure to provide that information constitutes a breach of the authorization, and the Environment Agency can take enforcement action. At the beginning of each calendar year, operators are required to submit details of the quantities of substances released during the previous year. This information is validated by the relevant Environment Agency inspector, and the information is passed for inclusion in the Chemical Release Inventory (CRI) database.[23] These data are being used in compiling the urban emissions inventories.

The Environmental Protection Act has been introduced in stages. Fuel production and combustion processes, including electricity generation, and waste disposal came under control in 1992. The mineral industries and chemical industries (excluding inorganic processes and chemical storage) came under control in 1993, and were followed by the remainder of the chemical industries in 1994. The metal industries were brought under control during 1995, and were followed during 1996 by the 'other industries' sector. The metal industries are not required to provide information on the annual amounts of substances released until 1997, and the 'other industries' sector until 1998, but are required to undertake routine emissions monitoring from the point at which authorizations came into effect.

Part B processes, designated for local authority control, must not operate without an authorization from the local authority. Under transitional arrangements, existing processes may continue to operate prior to receiving (or being refused) an authorization. Operators must submit a detailed application for authorization to the local authority. The local authority is obliged to include conditions in any authorization in order to ensure that the process is operated using the Best Available Techniques Not Entailing Excessive Cost (BATNEEC).

Process guidance notes were prepared by the DoE and published between 1991 and 1995 covering all the main categories of processes falling under local authority air pollution control. The notes, which are subject to a four-year review, give guidance on the types of emissions which are likely to arise as a result of each process, and set limits for emission concentrations.

Whilst this system of control provides a great deal of information on Part B industrial processes and their location, it does not generally provide the information which is required to compile emissions inventories. In most cases authorizations specify release concentrations, and any monitoring relates to these rather than total annual emissions. Unlike Part A processes, there is no requirement for operators of processes to submit details of the quantities of substances released during the preceding year. The DoE was recommended in the West Midlands report to consider bringing the reporting procedures for Part B processes under local authority control into line with those for Part A processes.[24]

Whilst industrial processes result directly in emissions, in many cases the majority of the emissions occur as a result of the fuel use associated with the process. In addition to the industrial processes, information is collected on the boiler plants larger than 2 megawatts but not large enough to come within the

[23] Her Majesty's Inspectorate of Pollution, *The Chemical Release Inventory* 1994, HMSO, London, 1996.
[24] D. Hutchinson and L. Clewley, Ref. 12, p. 36.

ambit of the Environmental Protection Act 1990. This information also provides a valuable cross-check against the records of fuel supplied.

## 3 West Midlands Atmospheric Emissions Inventory

The West Midlands Atmospheric Emissions Inventory[25] is the first in the series of urban inventories being prepared under the DoE's Air and Environment Quality Research Programme to be completed. Inventories are also currently being prepared for Bristol, Portsmouth/Southampton, Glasgow, Liverpool/Manchester and Swansea/Neath/Port Talbot as part of the series of 10 further inventories being compiled over the next two years.

The West Midlands inventory includes the pollutants listed in Section 2 above. It does not include all pollutants which may be of concern for health or other reasons. Whilst airborne lead levels have fallen significantly since 1981, when the amount of lead in petrol was first reduced, further consideration is currently being given to industrial sources in the West Midlands. Lead and other pollutants may be included in future inventories.

The area covered by the seven West Midlands Metropolitan Councils (Birmingham, Coventry, Dudley, Sandwell, Solihull, Walsall and Wolverhampton) is not a single entity but consists of a multi-centred conurbation which physically coalesced in the period of rapid urban growth between the late 1940s and 1970s (Figure 4). During the Industrial Revolution, the presence of large quantities of coal, limestone, clay, sand and gravel enabled the establishment of the heavy industries such as steel, iron, brick and glassworks in the Black Country area. These industries developed within villages, each with its residential, commercial and industrial areas, which then expanded into each other.

As can be seen from Table 2, the single most significant source of atmospheric pollutants in the West Midlands is road traffic. In the case of CO, benzene and buta-1,3-diene, road traffic accounts for over 96% of emissions. This is because these pollutants are particularly associated with petrol-fuelled combustion engines, used in cars, and have few other significant sources. Road traffic also accounts for 85% of emissions of $NO_x$ and 75% of black smoke, but only 17% of $SO_2$. $SO_2$ emissions have been reduced very substantially over the past 30 years. Nevertheless, industrial combustion remains the principal source of $SO_2$ emissions in the West Midlands.

Industrial processes are also a significant source of non-methane volatile organic compounds (NMVOC). These processes comprise metal and other coating processes, including car manufacturing and re-spraying, printing, the manufacture of paints and other coating materials, and the use of cleaning agents. These industrial sources, together with household use of paints, solvents, cleaning agents and cosmetics, are similar in scale to NMVOC emissions from vehicles, including evaporative emissions.

Black smoke has been included in this inventory as one of three measures of particulate emissions because of the large volume of historic data relating to black smoke. The single most significant source (67%) is diesel vehicles. Diesel

[25] D. Hutchinson and L. Clewley, Ref. 12, p. 00.

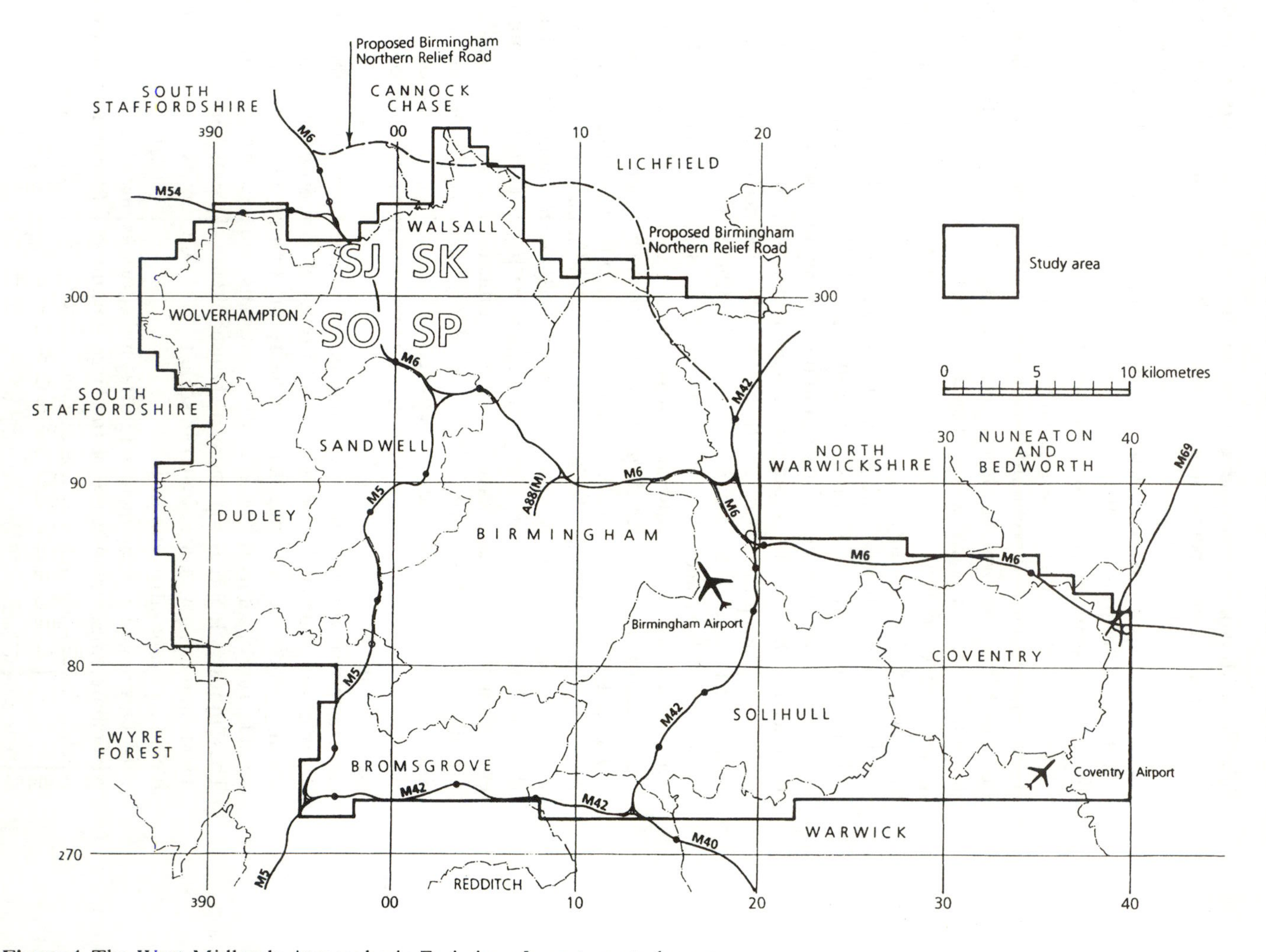


**Figure 4** The West Midlands Atmospheric Emissions Inventory study area

**Table 2** Estimated emissions in the West Midlands 1994 (percentages)

| *UNECE source category* | *Sulfur dioxide* | *Nitrogen oxides* | *Carbon monoxide* | *Carbon dioxide (as carbon)* | *Methane* | *Volatile organic compounds* | *Benzene* | *Buta-1,3-diene* | *Black smoke* | *Total suspended particulates* | $PM_{10}$ |
|---|---|---|---|---|---|---|---|---|---|---|---|
| 1 Public power, cogeneration and district heating | 0 | 0 | 0 | 0 | 0 | 0 | 0 | 0 | 0 | 0 | 0 |
| 2 Commercial, institutional and residential combustion | 13 | 6 | 1 | 40 | 0 | 0 | 0 | 0 | 1 | 10 | 8 |
| 3 Industrial combustion | 48 | 3 | 0 | 11 | 0 | 0 | 0 | 0 | 2 | 30 | 27 |
| 4 Non-combustion industrial processes | 15 | 1 | 0 | 1 | 0 | 0 | 0 | 0 | 1 | 3 | 0 |
| 5 Extraction and distribution of fuels | 0 | 0 | 0 | 0 | 10 | 2 | 0 | 0 | 0 | 0 | 0 |
| 6 Solvent use | 0 | 0 | 0 | 0 | 0 | 49 | 0 | 0 | 0 | 0 | 0 |
| 7 Road transport | 17 | 85 | 98 | 43 | 1 | 46 | 99 | 97 | 75 | 43 | 56 |
| 8 Other transport | 3 | 2 | 0 | 4 | 0 | 0 | 0 | 3 | 16 | 1 | 1 |
| 9 Waste treatment and disposal | 5 | 3 | 0 | 1 | 87 | 1 | 0 | 0 | 5 | 14 | 7 |
| 10 Agriculture | 0 | 0 | 0 | 0 | 2 | 0 | 0 | 0 | 0 | 0 | 0 |
| 11 Nature | 0 | 0 | 0 | 0 | 0 | 1 | 0 | 0 | 0 | 0 | 0 |
| TOTAL (tonnes) | 10 109 | 46 519 | 159 300 | 7 581 483 | 103 286 | 78 580 | 8484 | 172 | 8169 | 5491 | 4021 |

Source: West Midlands Atmospheric Emissions Inventory.
Figures are rounded and do not necessarily add to 100.

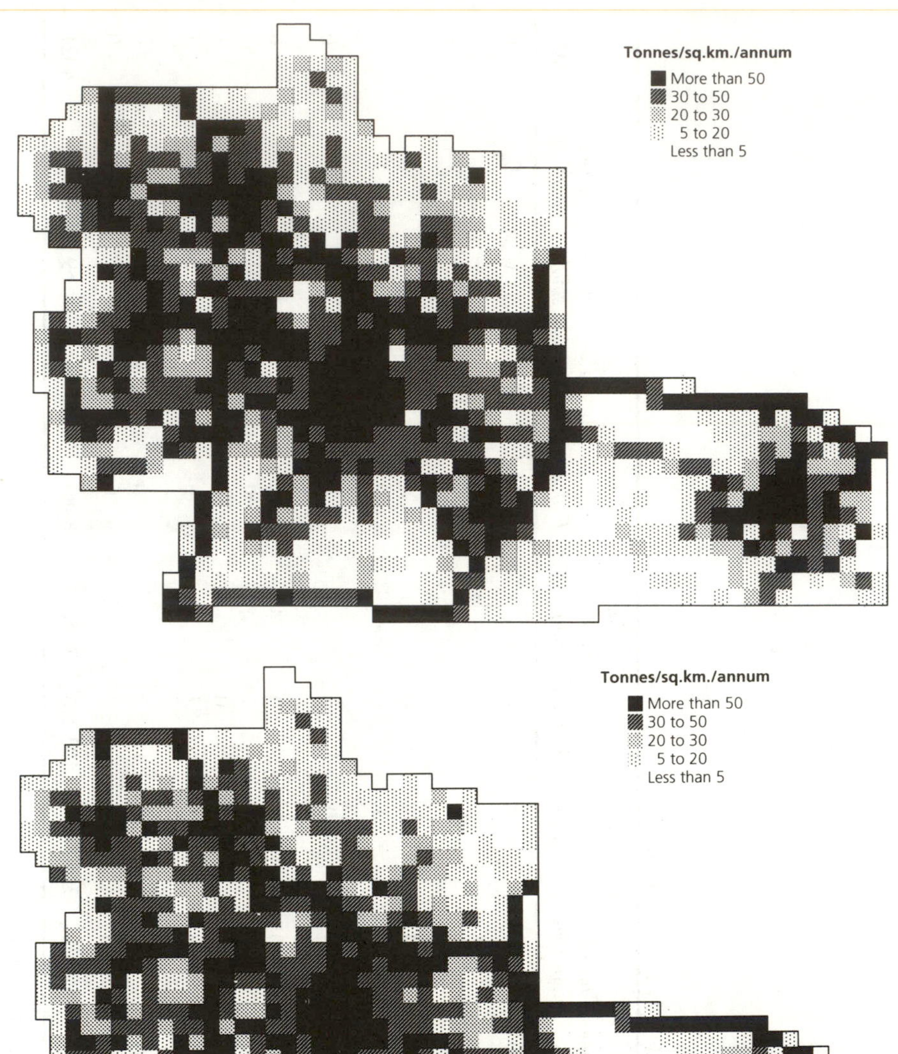


**Figure 5** Emissions of oxides of nitrogen from all sources in the West Midlands

**Figure 6** Emissions of oxides of nitrogen from road traffic in the West Midlands

vehicles are also a significant source of TSP (28%) and $PM_{10}$ (39%). Virtually all emissions of particulates from diesel vehicle are less than 10 $\mu$m in aerodynamic diameter.

The intensity of emissions varies considerably across the West Midlands, depending on the sources. This variation is illustrated in a series of maps which show the distribution of the emissions of each pollutant. The maps showing the distribution of $NO_x$ emissions from road traffic and from all sources are reproduced here as Figures 5 and 6. They clearly show the lines of the M6 and other motorways and main roads as well as the concentrations of traffic in central Birmingham, Coventry and other town centres.

## 4 Conclusions

Emissions of smoke and $SO_2$ have fallen dramatically since they were the cause of the great London smog in 1952. However, air quality is again causing concern at the international, national and local level, particularly as a result of emissions from motor vehicles. Although the present situation is less serious than that in the 1950s, results from the automatic air quality monitoring network show that concentrations of pollutants exceed non-mandatory health guidelines in many British cities. The preparation of emissions inventories is essential to understanding the source of air pollution, and a prerequisite to the taking of effective action to improve air quality.

This review describes procedures which are still being developed. The various international, national and local projects do not have the resources to collect vast amounts of new data and must often rely on information which is already available and, in many cases, has been collected for other purposes. Space does not allow for a full discussion of each level of inventory and the review has therefore focused on the procedures being used to prepare urban inventories in the UK. The methodology described seeks to achieve a reasonable balance between accuracy and cost-effectiveness in delivering what is required. The cross checks carried out, for example on road transport fuel use, suggest that it produces relatively accurate results. Even so, the methodology will continue to be improved and refined.

*Postscript*

Since this chapter was written, the Government has published the definitive version of *The United Kingdom National Air Quality Strategy*, and the Department of the Environment (DoE) has become the Department of the Environment, Transport and the Regions (DETR).

# Ambient Air Quality Monitoring

JON BOWER

## 1 Introduction

Concern is increasing world-wide about the effects of air pollution on health and the environment. Increasingly rigorous legislation, combined with powerful societal pressures, are increasing our need for impartial and authoritative information on the quality of the air we all breathe. Monitoring is a powerful tool for identifying and tackling air quality problems, but its utility is increased when used, in conjunction with predictive modelling and emission assessments, as part of an integrated approach to air quality management.

Whatever your reason for monitoring—whether to meet local or national objectives, assessing traffic or industrial impacts, planning, policy development or providing public information—your measurements will need to be accurate and reliable if they are to prove useful. This is why proper quality assurance and control (QA/QC) is a key component of any monitoring programme. Without this, your data will not provide a sound basis for decision making or air quality management; as a result, any investment of money, time and effort made in monitoring will have been wasted.

In its fullest sense, a QA/QC programme should cover all aspects of network operation, ranging from system design and site selection through equipment selection, operation, calibration and maintenance to data management and validation.

The importance of clearly defined objectives, as the first step to designing or implementing any air monitoring system, is emphasized in this article. Basic network design considerations are reviewed. Site numbers, locations and sampling strategies will be determined by overall monitoring objectives and resource availability. Establishing management and training systems also forms an important component of the network design phase.

The selection of appropriate and cost-effective monitoring methodologies is vital. Trade-offs between equipment cost, complexity, reliability and performance are examined; the use of simple technologies, where applicable, is recommended. System operation, calibration and maintenance activities are considered. Pro-active site auditing, and the establishment of a direct calibration-chain to reference gas standards, are shown to be important network operational requirements.

Effective data management, screening and validation is always necessary to ensure robust and credible network outputs. It should be recognized that raw data, in isolation, will usually be of limited utility; these need to be appropriately analysed and interpreted in order to provide useful information for technical, policy and public end-users. This review aims to provide a basic, step-by-step and practically based approach to establishing and operating cost-effective ambient air quality monitoring programmes.

## 2 The Role of Monitoring in Air Quality Management

Air quality management is fundamentally concerned with the achievement of economic, awareness and regulatory objectives. In order to develop or implement an effective air quality management (AQM) plan, it is first necessary to obtain reliable information on ambient pollution levels. This point was fully recognized in Agenda 21 of the United Nations Conference on Environment and Development (UNCED), held in Rio de Janeiro in 1992.[1]

The ultimate purpose of monitoring is not merely to collect data (a common perception!) but to provide the necessary information required by scientists, policy makers and planners to enable them to make informed decisions on managing and improving the environment. Monitoring fulfils a control role in this process, providing the necessary sound scientific basis for policy and strategy development, objective setting, compliance measurement against targets, and enforcement action.

However, the limitations of monitoring should also be recognized. In many circumstances, measurements alone may be insufficient for the purpose of AQM. Monitoring therefore often needs to be used in conjunction with other objective assessment techniques, including modelling, emission measurement and inventories, interpolation, mapping and interpretation. Some of these techniques are explored in greater detail in this review.

Ambient air quality measurements and modelling-based assessments can usefully be regarded as complementary activities. No monitoring programme, however well funded and designed, can hope to quantify patterns of air pollution comprehensively in both space and time. At best, monitoring provides an incomplete—but useful—picture of the environment. Conversely, reliance on modelling alone is equally unsound. Although models can provide a powerful tool for interpolation, prediction and optimization of control strategies, they are effectively useless unless properly validated by real-world monitoring data. Accordingly, ambient monitoring should be regarded as a component—albeit a key one—in any integrated approach to managing air quality.

## 3 Monitoring Objectives

We make no excuses for starting at this point. The first step in designing or implementing any monitoring system is to define its overall objectives. Moreover, in order for a monitoring programme to be effective within the framework of an

[1] UN, *Report of the United Nations Conference on Environment and Development (UNECD)*, Rio de Janeiro, 3–14 June 1992, A/CONF.151/26, United Nations, 1992, vols. 1–3.

overall environmental management system (EMS), it is vital that these objectives be clearly stated and understood. Setting diffuse, overly restrictive or ambitious monitoring objectives will result in cost-ineffective programmes with poor data utility. In such circumstances, it will not be possible to make optimal use of the available manpower and resources.

The relationships between the data collected and the information to be derived from it must be taken into account when a monitoring programme is planned, executed and reported. This emphasizes the need for users and potential users of the data to be involved in the planning of surveys, not only to ensure that they are appropriate to their needs, but also to justify the resource commitment. It is worth stressing that any organization is likely to have its own specific objectives for monitoring. These may be influenced by local or national requirements, as well as by international obligations.

There are many rationales for monitoring. These include statutory requirement, policy and strategy development, local or national planning, measurement against standards, identification/quantification of risk and public awareness. Typical monitoring functions are summarized in Box 1.

**Box 1** Some key monitoring objectives

- Identifying threats to natural ecosystems or population health
- Informing the public about air quality and raising awareness
- Determining compliance with national or international standards
- Providing objective inputs to AQM, traffic and land-use planning
- Policy development and prioritization of management actions
- Development/validation of management tools (models, GIS, *etc.*)
- Assessing point or area source impacts
- Trend qualification, to identify future problems or progress against management/control targets

Every monitoring survey or network is different, being influenced by a unique mix of local/national issues and objectives. Subsequent parts of this review tackle the problems of how to reconcile these often conflicting requirements. It is vital that clear, realistic and achievable monitoring objectives be set. This enables appropriate data quality objectives (DQOs) to be defined (Box 2). In turn, this makes it possible for a targeted and cost-effective quality assurance programme (QAP) to be developed, networks to be optimally designed, priority pollutants and measurement methods to be selected, and data management/reporting requirements to be identified (Figure 1).

## 4 Quality Assurance and Control

Quality assurance and control (QA/QC) is an essential part of any air monitoring system. It is a programme of activities which ensures that your measurements meet defined and appropriate standards of quality, with a stated level of confidence. It should be emphasized that the function of QA/QC is not to achieve

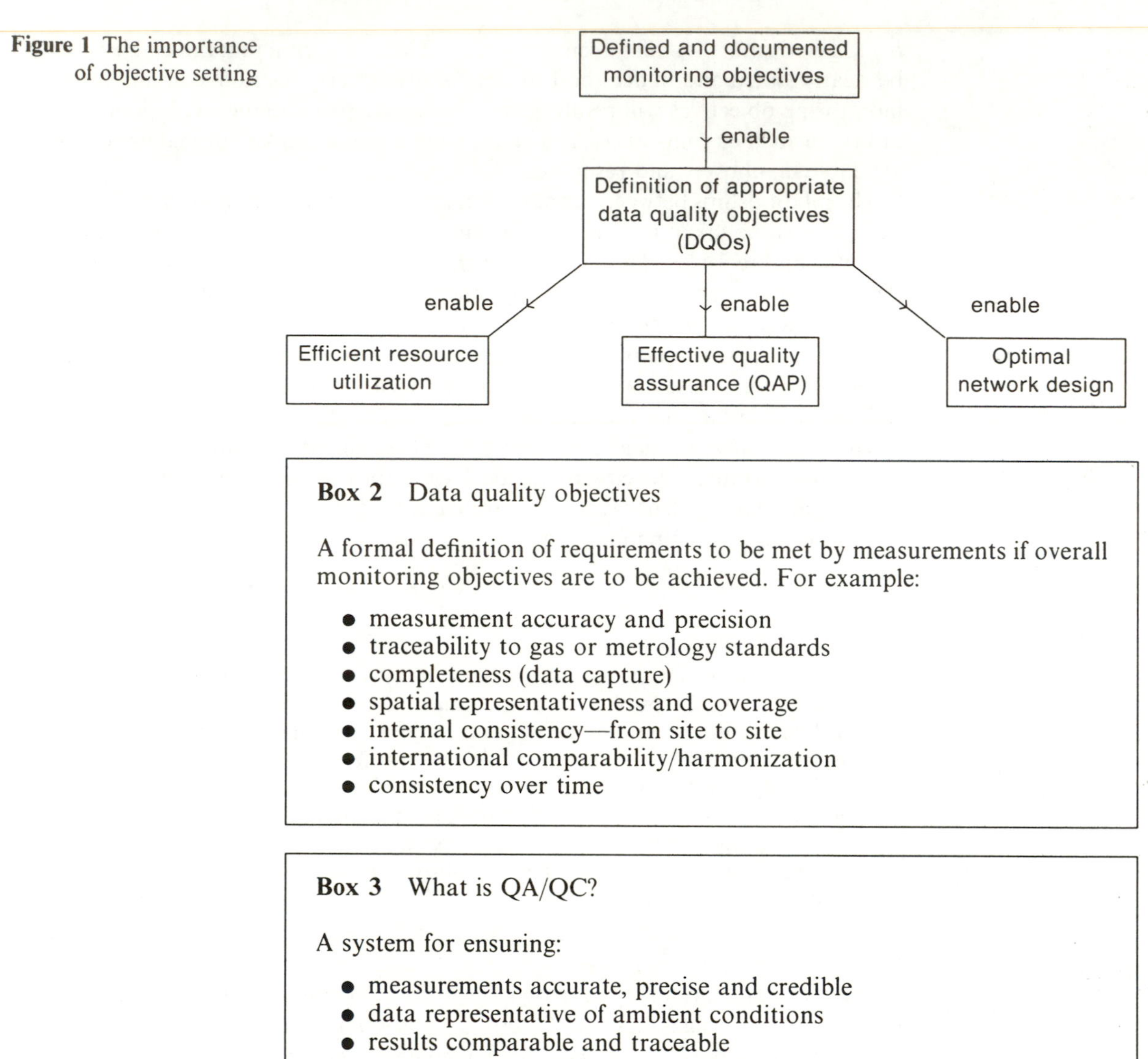


**Figure 1** The importance of objective setting

**Box 2** Data quality objectives

A formal definition of requirements to be met by measurements if overall monitoring objectives are to be achieved. For example:

- measurement accuracy and precision
- traceability to gas or metrology standards
- completeness (data capture)
- spatial representativeness and coverage
- internal consistency—from site to site
- international comparability/harmonization
- consistency over time

**Box 3** What is QA/QC?

A system for ensuring:

- measurements accurate, precise and credible
- data representative of ambient conditions
- results comparable and traceable
- measurements consistent over time
- high data capture, evenly distributed
- optimal use of resources

the highest possible data quality. This is an unrealistic objective, which cannot be achieved under practical resource constraints. Rather, it is a set of activities which ensures that your measurements comply with the specific DQOs for the monitoring programme. In other words, QA/QC ensures that your data are fit for purpose. Major QA/QC objectives are summarized in Box 3, whilst the functional components of a QA/QC programme are identified in Box 4.

Quality assurance activities cover all pre-measurement phases of monitoring, including determining monitoring and data quality objectives, system design, site selection, equipment evaluation and operator training. Quality control functions

**Box 4** QA/QC for air monitoring: the major components

Quality Assurance
- definition of monitoring objectives
- network design, management and training systems
- site selection and establishment
- equipment evaluation and selection

Quality Control
- routine site operations
- establishment of calibration/traceability chain
- system maintenance and support
- data review and management
- system review and development

affect directly measurement related activities such as site operation, calibration, data management, field audits and training. In other words, quality assurance relates to the measurement process, whilst quality control is concerned primarily with outputs.

You will notice in Box 4 that we have identified a final QA/QC activity, that of system review. This recognizes that networks, and the QA/QC systems that serve them, are not static: they evolve over time as objectives, resources and air pollution climates change (Box 5). For example, in many developed or rapidly

**Box 5** A reminder: networks and QA systems are not static!

Periodic review is required to assess:

- changes in monitoring objectives/priorities
- new priority pollutants and measurement methods
- changes in site conditions
- new local or national legislation
- changing emission patterns/sources
- changes in resource availability

developing temperate countries, national measurement programmes have had to shift from a historic emphasis on smoke and sulfur dioxide (arising from domestic coal combustion) to a completely different problem, that of primary, secondary and air toxic pollution from motor traffic in cities.

In its fullest sense, QA/QC covers all aspects of network operation ranging from system design and site selection through equipment selection, operation, maintenance, calibration and operator training to data review and validation. The successful implementation of each component of a QA/QC scheme is necessary to ensure the success of the complete programme, and thus of the overall monitoring effort. The design of an effective and targeted QA/QC programme is, however, only the first step in the process of quality management. The programme needs to be fully documented, and compliance with its procedures and requirements actively monitored.

An ongoing resource commitment to QA/QC is required in any monitoring survey or network. Typically, a budget of between 20% and 50% of the total annual operating costs may be appropriate for this purpose, depending on the complexity of the programme and the stringency of its DQOs.

In large-scale or multi-operator networks (such as the UK's national monitoring networks) or international measurement programmes (*e.g.* GEMS/Air or the emerging EURO-AIRNET), a common approach to QA/QC can be an important tool in maximizing data consistency and reliability. Formal accreditation systems, such as ISO 45001, may have an important role to play in future in harmonizing QA/QC and network operational systems over an international scale. A step-by-step model for QA/QC programmes is provided in Figure 2, whilst specific components are discussed further in subsequent sections.

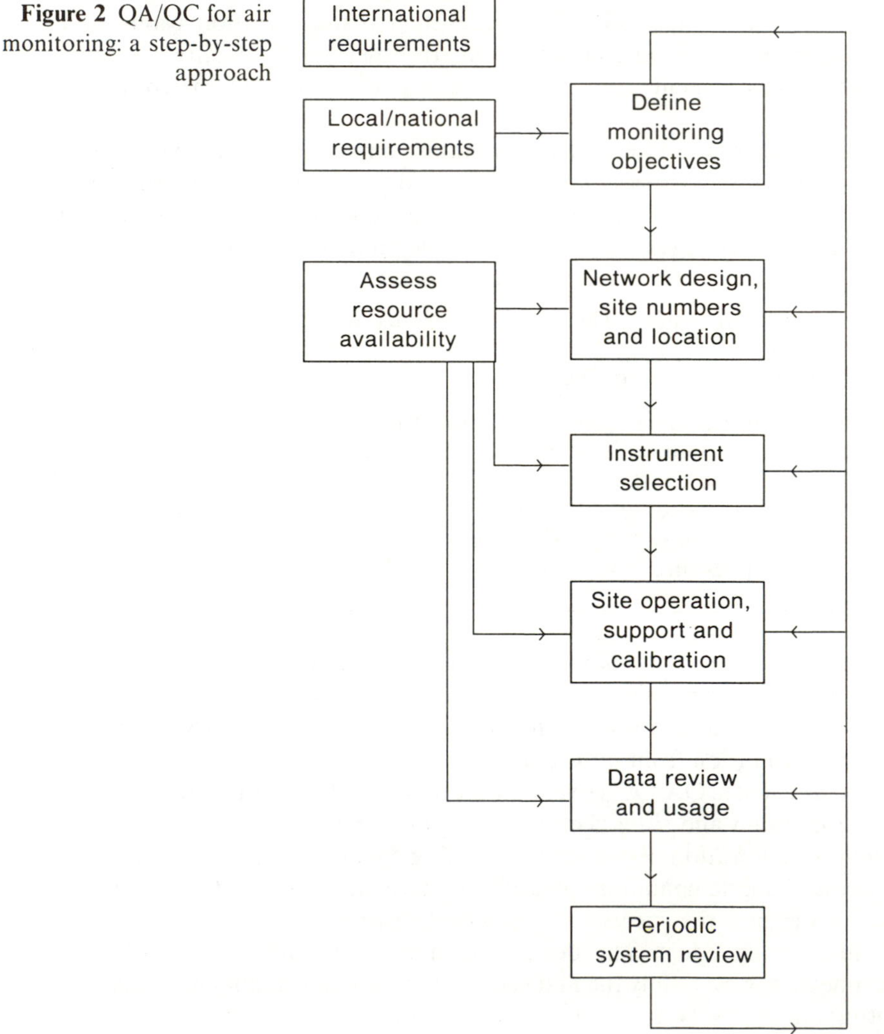


**Figure 2** QA/QC for air monitoring: a step-by-step approach

## 5 Network Design

### *Basic Design Features: Objectives and Resources*

There are no hard and fast rules for network design, since any decisions made will be determined ultimately by the overall monitoring objectives and resource availability. Although monitoring systems can have just a single, specific objective, it is more common for there to be a broad range of targeted programme functions. No survey design can hope to address completely all the possible monitoring objectives listed in Box 1. However, the design of surveys to meet these individual requirements often has common features, and can use common data (to avoid duplication of effort) and overlapping data to verify the credibility of results and conclusions. The overall design goal is to ensure that the maximum information can be derived from the minimum effort.

Another key issue, needing to be addressed at a very early stage, is that of resource availability (Box 6). In practice, this is usually the major determinant in network design, which will exert a particularly strong influence on the choice of site numbers, pollutants to be monitored and instrumentation selected. One should also be aware of the likely commitments and costs to be incurred in any air monitoring programme. Just some of these are listed in Box 7. Beware the dangers of spending before planning. Before any firm capital commitment, plan the survey, assess resource availability, select the most appropriate equipment and choose monitoring sites.

**Box 6** Network design: what are my resources?

- Money (capital and ongoing)
- Skilled manpower
- Time

**Box 7** Costs of air monitoring

- Capital purchase of analysers, ancillary equipment and site infrastructure
- Equipment service, maintenance and repair
- Staff and subcontractor costs—operational and management
- QA/QC—audits, intercalibrations, training, data management, *etc.*
- Running costs—site rental, electricity, consumables, spare parts, calibration gases, telephone, *etc.*

A very common problem, which we have seen manifested all over the world, is the purchase of expensive air pollution measurement equipment without regard to its long-term operational or financial sustainability. Local sustainability requires the continuing availability of agents (or an in-house capability) for repair and maintenance, together with the necessary skill-base for routine equipment operation and calibration. Financial sustainability recognizes the need for an ongoing budget for equipment operation, typically amounting to ~10% per annum of the initial capital expenditure.

## *Site Numbers and Selection*

In practice, the number and distribution of air quality monitoring stations required in any extended network depend on the area to be covered, the spatial variability of the pollutants being measured and the required data usage (Box 8).

**Box 8** Network design (i): site numbers

Will depend on:

- required data use/objectives
- area to be covered
- spatial variability of pollutants
- resource availability
- instruments deployed

There are a number of approaches to network design and site selection. Source-orientated monitoring often targets worst-case 'hot-spot' environments, whilst background or baseline studies are usually optimized for assessing general population or ecosystem exposure to pollutants of concern. Although the overall requirement is to maximize spatial coverage and representativeness, in practice this goal is only approached by grid-based monitoring strategies: these can be optimized to provide detailed information on spatial variability and exposure patterns for priority pollutants. However, this approach is highly resource-intensive and not, in consequence, widely used. To reduce resource requirements, a grid approach can be utilized in conjunction with intermittent or mobile sampling. However, use of this technique is not consistent with the need to maximize temporal representativeness as well as spatial coverage (see next subsection).

A more flexible approach to network design, appropriate over city-wide or national scales, involves siting monitoring stations or sampling points at carefully selected representative locations, chosen on the basis of required data usages and known emission/dispersion patterns of the pollutants under study. Some of the factors to be considered in site selection are detailed in Box 9. This approach to network design requires considerably fewer sites than grid strategies and is, in consequence, cheaper to implement. However, sites must be carefully selected if measured data are to be useful.

Microscale siting considerations are also important in ensuring that meaningful and representative measurements are made. If baseline concentrations are to be assessed, then monitoring sites should be adequately separated from local pollutant sources or sinks. Probe aerodynamics and site sheltering may also be important.

A variety of practical considerations apply when selecting monitoring sites. They must be accessible for site visits, but the potential for public interference or

**Box 9** Network design (ii): site locations

Factors to consider include:

- major pollutant sources/emissions in area
- target receptors and environments
- meteorology and topography
- model simulations of dispersion patterns in area
- existing air quality information (*e.g.* from screening studies)
- demographic/health/land-use data

vandalism must also be recognized. The availability of mains electricity for pollutant analysers and station infrastructure, together with a telephone linkage for data telemetry (if utilized) must be feasible (Box 10).

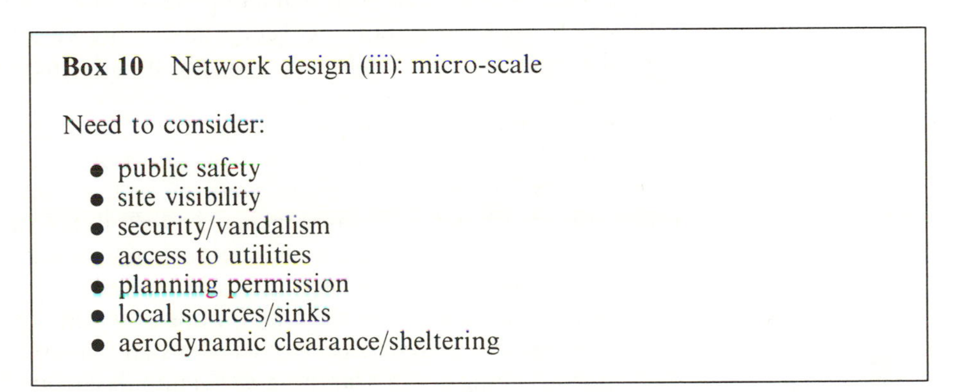

**Box 10** Network design (iii): micro-scale

Need to consider:

- public safety
- site visibility
- security/vandalism
- access to utilities
- planning permission
- local sources/sinks
- aerodynamic clearance/sheltering

Public safety is a paramount issue in site selection. The author had the misfortune, many years ago, to be responsible for a hydrocarbon measurement site employing an FID analyser that malfunctioned and caused a hydrogen explosion. Fortunately there was no injury, just considerable expense and embarrassment. We have not used this analyser type since! Other potential risks to be aware of may include the use of compressed gas cylinders for on-site calibrations, heavy roof-mounted equipment (*e.g.* High Volume particulate samplers) in typhoon or hurricane-prone areas, and even mundane issues such as site doors opening unexpectedly onto public thoroughfares.

## *Sampling Strategies and Systems*

Monitoring involves qualifying pollutant behaviour in both space and time. A good network design should therefore seek to optimize both spatial and temporal coverage, within available resource constraints.

The previous section dealt with the task of maximizing spatial coverage and representativeness of measurements. Achieving good time-domain performances is not a problem for most commonly used air monitoring methodologies:

continuously operating automatic analysers readily provide adequate temporal representativeness and resolution, as do well-recognized semi-automatic methods such as OECD-type smoke/$SO_2$ samplers.

Integrating measurement methods such as passive samplers, although fundamentally limited in their time resolution, still produce data which are representative of their total exposure period. Problems can arise, however, when using manual sampling methods in an intermittent, mobile or random deployment strategy. Such an approach is usually adopted for operational or instrumental reasons, or simply because it would not be possible to analyse the sample numbers produced by continuous operation.

Intermittent sampling is still surprisingly widely used, for instance in parts of Eastern Europe and China. The 'spot' measurement strategy ($\sim$15–20 minutes at a time) still mandated for sampler programmes in some parts of China involves total annualized data capture rates of less than 0.02%, and is therefore of limited utility in assessing diurnal, seasonal or annual pollutant patterns.

It is important to recognize that sampling strategies must provide data of sufficient temporal resolution, coverage and representativeness to meet the overall monitoring aspirations. Relatively few monitoring objectives can be met by intermittent or 'grab' sampling.

The importance of a properly designed air sampling system cannot be overemphasized. In order for automatic analysers or samplers to reliably measure ambient pollutant concentrations, it is essential that these pollutants are transferred unchanged to the instrument reaction cell. The sampling manifold is a crucial and often overlooked component of any monitoring system, which strongly influences the overall accuracy and credibility of all the measurements made. When auditing sites in the UK, Europe or world-wide, it is interesting to note that sampling system deficiencies—usually inappropriate designs or inadequate cleaning—are by far the most commonly encountered problem. Some design requirements common to all gas sampling systems for gaseous air analysers are summarized in Box 11.

**Box 11** Key sampling system design features

- Inertness to pollutants being sampled
- Minimized air residence time
- Low airstream/sample line interaction
- Excess flow above total analyser demand
- Minimized pressure drop
- Removal of interferents such as water vapour/pollutants
- Avoidance of thermal 'shock' when air sampled
- Ease of cleaning and maintenance

## *Management and Training Systems*

Adopting appropriate network management and organizational systems is important in ensuring the success of any air monitoring programme. Many UK and European networks are organized on an increasingly decentralized basis, with different organizations responsible for network management, quality control and site operations. Support functions such as equipment service and maintenance are usually subcontracted.

The two contracting approaches to the organization of large-scale networks may be exemplified by the UK and Dutch network monitoring programmes. The current 80-site UK urban network involves nearly as many organizations participating or 'stakeholding' in the programme. By contrast, the comparable Dutch programme is highly centralized, being operated and managed by one organization.

A devolved decentralized network structure will clearly not be appropriate for small-scale local or research networks, or, indeed, individual sites. For a large network, it offers obvious advantages and disadvantages. Decentralized networks can be effective in fostering 'trickle down' of expertise from experienced national laboratories to local site operators. However, this can generate a significant training commitment to ensure competent field operations. By spreading the workload, very large developed networks may be established without a correspondingly large central government commitment. However, such programmes can lack flexibility when responding to changed monitoring requirements, as a consensus approach to decision making is often mandated.

It should be recognized that multi-operator networks require additional QA/QC commitment in order to harmonize operations and ensure consistent measurements. They are also harder to accredit to ISO 45001 or other recognized schemes. However, the existence of a separate QA/QC unit can ensure transparent and genuinely independent quality management of the overall monitoring system. When decentralized networks become very large, a major and ongoing commitment is required to manage the organizational interfaces, thereby ensuring smooth, friction-free network operation. Without this, a fragmented structure with poorly harmonized outputs can result.

## 6 Instrument Selection

The simplest methods that meet your monitoring objectives should always be selected. Inappropriate, too complex or failure-prone equipment can result in poor network performance, limited data utility and—worst of all—a waste of your money. As a general rule, only proven and generally accepted measurement methods should be considered. Some general ground rules for equipment selection are summarized in Box 12.

**Box 12** Choosing the right instrument

- Consider monitoring objectives and DQOs
- Required time resolution of measurement
- Resource availability
- Talk to other users
- Independent type approval/designation

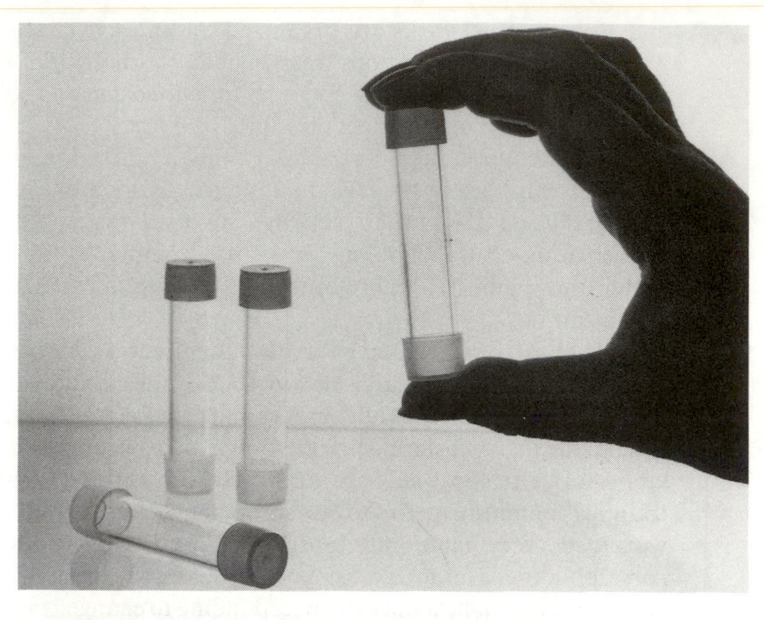

**Figure 3** Plate diffusion tube samplers: a simple and cost-effective way of measuring air quality

**Figure 4** Active samplers can still be a useful technique for many types of air survey

Air monitoring methodologies can be divided into four main types, covering a wide range of costs and performance levels (Figures 3–5). The main advantages and characteristics of these techniques are summarized in Table 1. Always choose the simplest technique that will do the job. Although your monitoring objectives are the major factor to consider, resource constraints and the availability of skilled manpower must also be considered. When budgeting for instrument purchase, do not forget to include ongoing costs for equipment support, calibration and maintenance in addition to capital purchase expenditure.

There is a clear trade-off between equipment cost, complexity, reliability and

**Figure 5** Remote sensors can be used for open-path, multi-pollutant measurements in real time

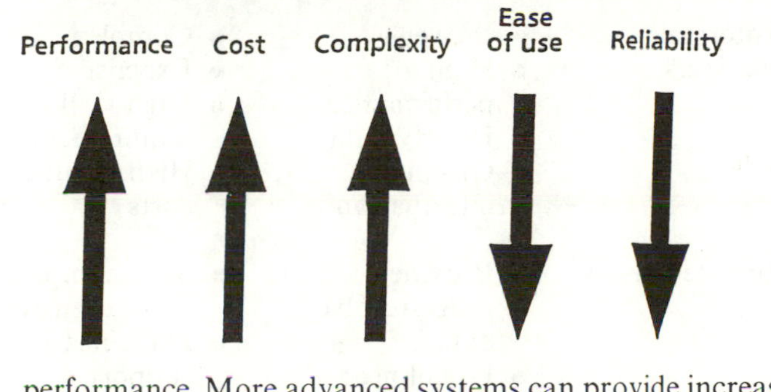


**Figure 6** Performance trade-offs for air monitoring

performance. More advanced systems can provide increasingly refined data, but are usually more complex and difficult to handle (Figure 6).

Do not discount active samplers such as bubblers just because higher technology measurement methods are available. Many baseline monitoring, area screening and site selection functions can be perfectly well served by such simple monitoring methodologies. Automatic analysers are expensive and not always straightforward to operate successfully. These should only be considered if your monitoring objectives have an explicit requirement for short-term measurements, if hourly measurements—or a particular automatic measurement methodology—are required by relevant legislation, or if rapid data dissemination and use is intended.

In practice, the combined use of samplers and automatic analysers in a 'hybrid' monitoring programme can offer a versatile and cost-effective approach to network design over a municipal or national scale. Such a network design will use passive or active samplers to provide good spatial coverage and resolution of measurements. Automatic analysers, deployed at carefully selected locations, can provide more detailed time-resolved data for assessing peak concentrations or comparison with short-term standards.

Remote sensors such as OPSIS can provide integrated multi-pollutant measurements along a specified path. This is the most expensive type of air

**Table 1** Air monitoring techniques

| Method | Advantages | Disadvantages | Capital cost |
|---|---|---|---|
| Passive samplers | • Very low cost<br>• Very simple<br>• Useful for screening and baseline studies | • Unproven for some pollutants<br>• In general only provide monthly and weekly averages | £8–50 per sample |
| Active samplers | • Low cost<br>• Easy to operate<br>• Reliable operation/ performance<br>• Historical dataset | • Provide daily averages<br>• Labour intensive<br>• Laboratory analysis required | ~£1k per unit |
| Automatic analysers | • Proven<br>• High performance<br>• Hourly data<br>• On-line information | • Complex<br>• Expensive<br>• High skill requirement<br>• High recurrent costs | ~£10k per analyser |
| Remote sensors | • Provide path or range-resolved data<br>• Useful near sources<br>• Multi-component measurements | • Very complex and expensive<br>• Difficult to support, operate, calibrate and validate<br>• Not readily comparable with point measurements | ~£50–100k per sensor, or more |

monitoring system, and careful attention needs to be paid to calibration, quality control and data validation procedures if the best use is to be made of this technology.

The European Framework Directive on air quality assessment and management explicitly recognizes the utility of using a broad range of monitoring, modelling and objective estimation techniques for assessing air quality in Member States, the technique of choice being dependent on the air quality status of the area under study.[2]

Any ambient air quality analyser requires careful selection. An assessment of equipment specifications alone will not be adequate for this purpose. Some

[2] Council of the European Union, *Council Directive on Ambient Air Quality Assessment and Management*, FN 9040/96, *Official Journal of the European Communities*, 1996.

instrumental parameters are readily quantified and therefore often quoted by manufacturers; these include specifications such as linearity and response time, accuracy and precision, noise, drift and detection threshold/range. Although these factors are clearly important, they do not tell the whole story. Other instrumental characteristics, though less readily defined, may be of paramount importance in a practical context; these include reliability and ease of operation, compatibility with existing equipment, maintenance/calibration requirements and—especially in extended networks—the capability for unattended operation over long periods (Box 13).

**Box 13** Choosing analysers: what really matters?

- Suitability for application
- Simplicity and ease of use
- Reliability
- Cost of purchase and operation
- Proven performance
- After-sales support
- Local sustainability

## 7 System Operation

### *Site visits*

Frequent documented site visits are an essential component of any QA/QC system for air monitoring, although the frequency of visits required will vary from network to network. Telemetry systems can provide an efficient and cost-effective method for data acquisition from sites, but their adoption does not obviate the need for regular visits by operators. These should, in fact, be performed as frequently as operational needs, geographical constraints and available manpower permit.

It is a common oversight to place too much emphasis on laboratory-based QA/QC activities, as these are often easier to control, document and monitor. Although such activities are important, particularly for sampler-based measurement programmes involving laboratory sample analysis, considerable emphasis in any quality system needs to be focused on the point of measurement. Mistakes or problems at the start of the measurement chain cannot readily be corrected afterwards!

Many operations essential to maximize data integrity and capture rate must be carried out on-site. These are summarized in Box 14. To enable these functions to be carried out in an efficient and systematic way, a site visit schedule should be drawn up which makes provision for all monitoring sites to be visited regularly at specified intervals, typically between weekly and monthly. A comprehensive calibration record and instrument checklist should be completed after each site visit and retained for subsequent QA/QC checking.

**Box 14** Site visit functions

- Weekly to monthly frequency
- Ensure smooth running of equipment
- Calibration and diagnostic checks
- Anticipating future problems
- Change filters and consumables
- Check sampling system and pumps
- Cleaning of sampling systems
- Install/replace/repair equipment
- Check external site conditions

## *Equipment Support and Maintenance*

The importance of maintenance procedures for air quality analysers cannot be over-emphasized. It is only through proper instrument support that monitoring systems can be relied on to operate satisfactorily and for extended periods in the field. Maintenance schedules for the replacement of consumable parts, diagnostic checks and equipment overhaul should in all cases follow manufacturers' recommendations. Call-out procedures and repair turnaround times also need to be specified, in the event of equipment failure in the field. In many networks, equipment maintenance and support is subcontracted to local equipment service agents.

When considering the use of complex air monitoring technologies, it is important to recognize the need for resource commitments well beyond the initial capital investment costs. On-going expenditure will be required for the lifetime of the equipment to support the monitoring effort, maintain the equipment in an operational state, and ensure that meaningful data are being acquired.

## *Equipment Calibration*

Proper calibration of automatic monitoring equipment is essential for obtaining accurate and traceable air quality data. For most common gaseous air pollutants, this involves the use of on-site transfer gas cylinders or permeation sources to generate a reproducible equipment 'span point', thereby determining the system response to an accurately predetermined concentration of the air pollutant under analysis. An additional determination of the corresponding 'zero point', or system response when no pollutant is present (a measurement made using zero gas cylinders or suitably scrubbed air), suffices to give a 'two-point' calibration which is adequate for many purposes. 'Multipoint' calibrations involving several different span concentrations will be required in some circumstances, for instance after equipment servicing/repair or if linearity problems are suspected.

Manufacturers' or agents' own cylinder or permeation tube determinations cannot always be relied on for field calibrations; these sources should, wherever possible, be independently verified in the laboratory before use on-site. It is also necessary for them to be checked frequently during their operational lifetime, to identify drift or degradation.

The frequency and type of field calibrations required for any analyser should be defined in the quality assurance plan for the network. A typical scheme would include automatic calibration every 24 hours, using on-site permeation tube ovens or gas cylinders, and manual calibration using independent sources performed during every site visit. Regular intercalibrations involving all the network's analysers should also be performed (see below) in large networks.

Rigorously certified gas mixtures, or sources produced in-house, should always be used as primary laboratory references for the field transfer standards. Such primary standards must be directly traceable to absolute measurements or to accepted national or international metrology standards (Figure 7). A number of proven laboratory-based techniques are available for the preparation or verification of primary gas standards (Table 2). In practice, it is often desirable to prepare gases with one technique and verify or cross-check with others.

The importance of a sound primary calibration-base, and a clear traceability chain for all measurements, cannot be over-emphasized. These are fundamental determinants of measurement quality in any monitoring network.

## *Intercalibrations and Audits*

In UK's large-scale national networks, on-site calibration procedures are supplemented by regular audits and intercalibrations (Box 15), carried out by independent QA/QC Units. In-house teams or suitable contractors could also be used (Figure 8). Audits are typically arranged at least once a year. These provide an opportunity for a direct and qualitative assessment of operator procedures, site performance, infrastructure and instruments. They also allow data or instrument anomalies to be investigated on-site.

Intercalibrations may be performed every three to six months, depending on

**Figure 7** Establishing a sound and traceable gas calibration-base is vital to any programme

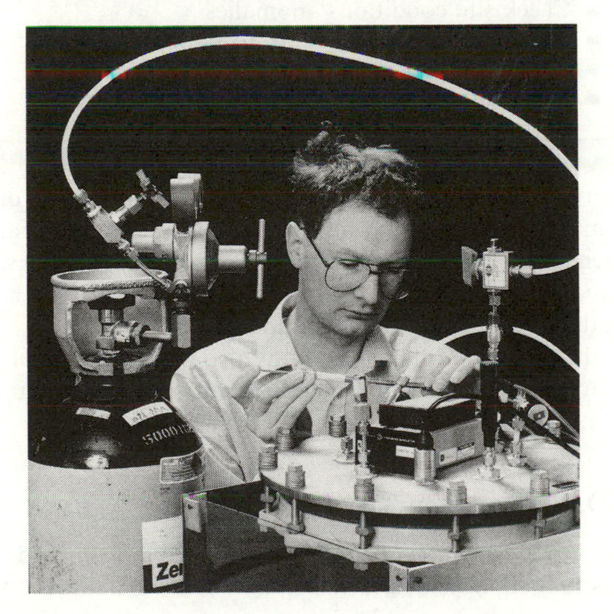

**Table 2** Primary gas calibration methods and traceability*

| *Method/pollutant* | *CO* | $SO_2$ | *NO* | $NO_2$ | $O_3$ | *Comments on method traceability* |
|---|---|---|---|---|---|---|
| Commercial cylinder | √ | — | — | — | — | Concentrations not assumed; must be checked by independent methods as appropriate |
| Permeation tubes | — | √ | — | √ | — | Absolute (weighing); commercial tubes may be traceable to standards |
| Static dilution | √ | √ | √ | √ | — | Absolute method (volume) |
| Dynamic dilution | √ | √ | √ | √ | — | Dependent on cylinder and mass flow controller performance |
| Gas phase titration | — | — | √ | — | √ | Not absolute but comparative technique ($O_3$/NO) |
| UV photometry | — | — | — | — | √ | Absolute method (UV absorption) |

√ appropriate method; — not applicable.

**Box 15** Site audits/intercalibrations

- Usually performed by external organization
- Ensure data comparability within networks
- Check site conditions/anomalies
- Establish traceability chain to national standards
- Check consistency of site operations
- Investigate systemic measurement anomalies

the network type. These involve a quantitative assessment of the full measurement system at each site, using common reference standards transported sequentially to each site in a network. In large-scale national networks, intercalibrations are essential to ensure data comparability and establish a direct measurement traceability chain to primary standards.

Both audits and intercalibrations also offer a powerful tool for harmonizing network performances and measurements across large international networks such as GEMS/Air or EURO-AIRNET.

## 8 Data Management

Even if all network operational guidelines described thus far have been successfully implemented, further quality control/assurance measures are necessary

**Figure 8** Pro-active audits provide a powerful tool for harmonizing network performance

in order to maximize data integrity. In any air monitoring network—however well implemented or operated—equipment malfunction, human error, power failures, interference and a wide variety of other disturbances can result in the acquisition of spurious or incorrect data; these must be filtered out before a final, definitive, database can be generated or used.

For convenience, the data review process is often regarded a two-stage process, of data validation followed by ratification. Data validation involves a rapid front-end screening process to identify or remove clearly spurious data prior to initial use; ratification refers to a long-term review of databases prior to final archival, analysis and reporting. In practice, the distinction is artificial and not widely recognized outside the UK. The review of measurements is best regarded as an integrated and ongoing data management activity.

## *Data Validation*

Rapid front-end screening of measurements is particularly important in networks with a commitment to real-time data dissemination to public or technical end users. However, it should be recognized that the screened data disseminated in this way are provisional, and may be subject to change as a result of subsequent ratification (see below). Some basic ground rules for data validation are reviewed in Box 16.

Many commercially available data telemetry and turnkey monitoring systems allow for the identification of out-of-range or suspect data and calibration factors. However, adherence to rigid data acceptance criteria and the automatic rejection of flagged data do not necessarily guarantee high data quality. For instance, a common consequence of following such pre-established data rejection routines is the invalidation of extreme (but valid) pollutant measurements simply

**Box 16** Data validation: some ground rules

- An on-going front-end screening process
- Review all data
- Do it quickly
- Use listings and graphs
- Common sense and experience required
- Avoid excessive dependence on automatic systems

because they lie outside pre-set or accepted limit values. Although leading-edge software tools such as neural networks offer promise of reducing routine data validation workloads, they can only be regarded as a useful tool rather than a complete solution at the present time.

Active examination of data throughput by skilled personnel can provide a more flexible approach to data validation. This review process is an important component of network quality assurance programmes; it serves both to identify possibly erroneous or invalid data and to inform field operators in good time of any equipment malfunction or problem requiring attention. Both daily data summaries and regular compilations (monthly and/or seasonal) of graphical data and calibration control charts may be used to assist front-end data review. The experience, common-sense and initiative of data-screening staff are prerequisites for the review process to be implemented successfully.

## *Data Ratification*

Unlike data validation, which is normally performed by site operators or network managers, data ratification for UK national networks is performed by independent QA/QC units. However, the use of in-house teams or suitable outside contractors would also be possible. Typically, the ratification process involves a quality circle of field operational, calibration laboratory, instrumentation and data management personnel.

Data ratification is not a mechanistic process, and does not readily lend itself to automation. Although software-based expert systems may be able to assist in future, there is no reliable substitute at present for the use of human judgement. Ratification is a high-skill exercise involving considerable knowledge of pollutant behaviour and dispersion, instrument characteristics, field experience and judgement (Box 17). Some of the many inputs to the ratification of national network datasets are summarized in Figure 9.

It may be noted that one of the major inputs to the ratification process is often the regular intercalibration dataset. Ratification is therefore typically based on three- to six-monthly databases, allowing long-term performance drift, site and instrumental anomalies to be reliably identified. Less frequent data ratification is not usually recommended; this may allow errors to propagate or worsen, leading to long-term data invalidity and rejection.

**Box 17** Data ratification

- A periodic, often 3–6 month, review
- The final stage of data acceptance
- Usually carried out by separate QA/QC unit
- Judgement and experience-based
- Many inputs

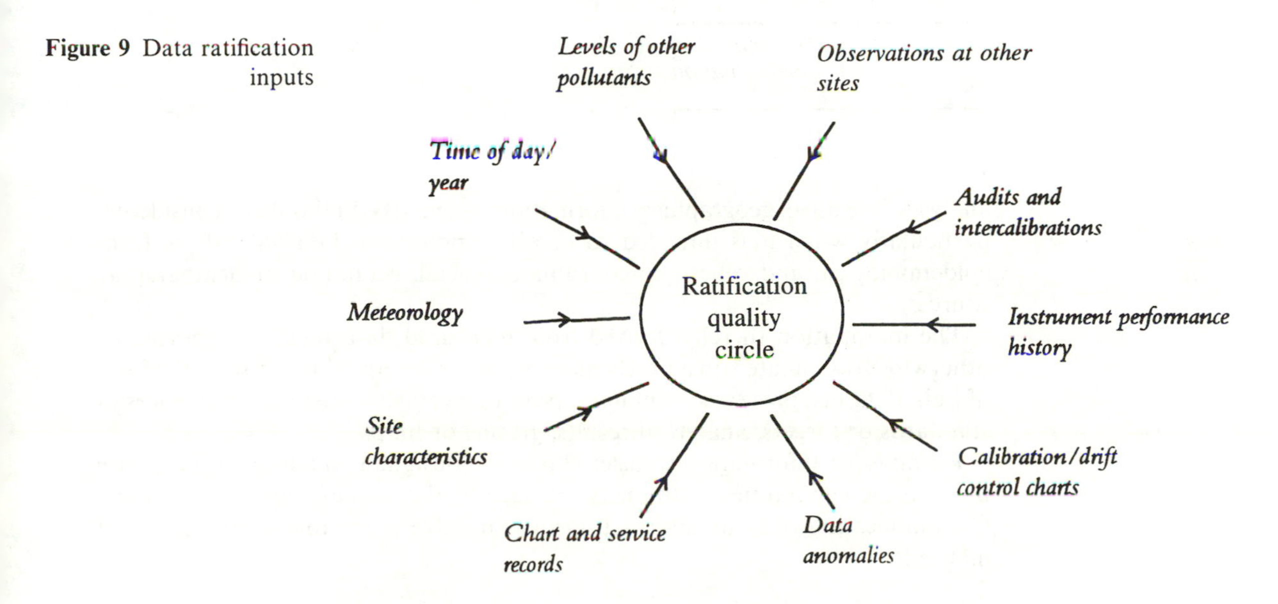


**Figure 9** Data ratification inputs

## *Turning Data into Information*

At the start of this review, we emphasized that the purpose of monitoring was not merely to collect data, but to produce useful information for technical, policy and public end-users (Figure 10). Raw data, in themselves, are of very limited utility. As we have seen, these first need to be screened (by validation and ratification) and collated to produce a reliable and credible dataset. In typical environmental management information systems (EMIS), the ratified measurements are databased together with corresponding emission datasets, model predictions and other relevant input to decision making. The next stage in data management is appropriate analysis and interpretation, designed to provide useful information—in an appropriate format—for end-users. A variety of proven analytical methodologies are available for air quality datasets.[3] In the final analysis, however, the appropriate level and method of data treatment will be very much determined by the ultimate end-use.

A minimum level of data management could be regarded as the production of daily, monthly and annual summaries, involving simple statistical and graphical

[3] WHO/UNEP, *GEMS: Analysing and Interpreting Air Monitoring Data*, World Health Organisation, Geneva, 1980.

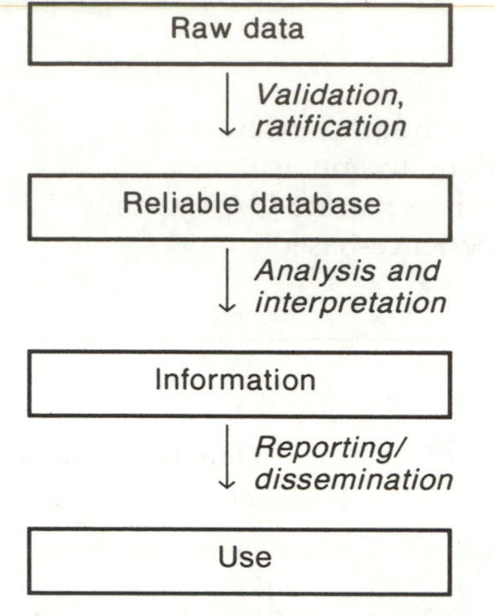


**Figure 10** Data throughput from a monitoring programme

analyses. The use of geographical information systems (GIS) should be considered, particularly when it is intended to combine pollution data with those from epidemiological and other geo-coordinated social, economic or demographic sources.

The information thereby derived from measured data must be reported or otherwise disseminated, in a timely manner, to end-users. This can be in the form of bulk datasets, processed summary, peak or average statistics, exceedences of standards or targets, analytical results, graphs or maps.

Formats for information transfer should be designed which are appropriate both to the capabilities of the network and to the requirements of the users. Communicating data or information may involve a number of transmission methods:

1. Paper. The longest-established communication method for air quality information is, of course, the written paper report. In the UK, for instance, comprehensive annual air quality summary reports are produced for both automatic and sampler networks,[4–6] together with information brochures for local authorities and the general public[7] and the deliberations of expert groups.[8] Some would argue, however, that this method is rapidly becoming obsolete in the light of advances in electronic data communication methods.
2. Computer Media. Based for many years on the transfer of floppy discs, this method has now been rapidly superseded in the UK by the use of on-line or broadcast methods, together with the annual publication of CD-ROMs containing both data and statistics from all national air monitoring programmes.

[4] J.S. Bower, G.F. Broughton, *et al., Air Pollution in the UK, 1995*. AEA Report No. AEA/RAMP/20112002/02, AEA Technology, Culham, 1997.

[5] A. Loader, K.J. Stevenson, *et al., UK Smoke and Sulphur Dioxide Monitoring Networks, 1995/6*, AEA Report No. 0911, AEA Technology, Culham, 1997.

[6] K.J. Stevenson and T. Bush. *UK Nitrogen Dioxide Survey, 1995*. AEA Report No. AEAT 0912, AEA Technology, Culham, 1996.

[7] H. Clark, *Air Quality in the UK*, NETCEN/Department of the Environment, London, 1996.

[8] Quality of Urban Air Review Group, *Urban Air Quality in the UK*, HMSO, London, 1993.

3. Electronic Media. These are, clearly, going to be the dominant communication medium in the future. For public and technical users in the UK, they are already the easiest and most accessible source of information on air quality.

Hourly updated data on all pollutants monitored in UK national automatic networks are available on CEEFAX and TELETEXT, weather bulletins and a free telephone service (Figures 11–13). These hourly data, together with ratified datasets, analyses, maps and information on current air quality issues are also

**Figure 11** During UK episodes, air quality warnings are broadcast on TV weather bulletins

P416 CEEFAX 416 Thu 15 Aug 15:38/47

BBC AIR 1/4

OZONE 1pm Thu 15 Aug
Southern England

| | Air Quality | Level (ppb) |
|---|---|---|
| **URBAN sites** | | |
| Bristol Centre | Very Good | 13 |
| S'hampton Centre | Very Good | 23 |
| Exeter Roadside | Very Good | 43 |
| **RURAL sites** | | |
| East Anglia | - | - |
| Oxfordshire | Very Good | 39 |
| Rochester | Very Good | 22 |
| South Somerset | Good | 54 |
| East Sussex | Good | 55 |
| Devon | Good | 60 |

Data from Department of Environment
London Air Qual Weather Travel

**Figure 12** Real-time information from all UK automatic monitoring networks is widely available

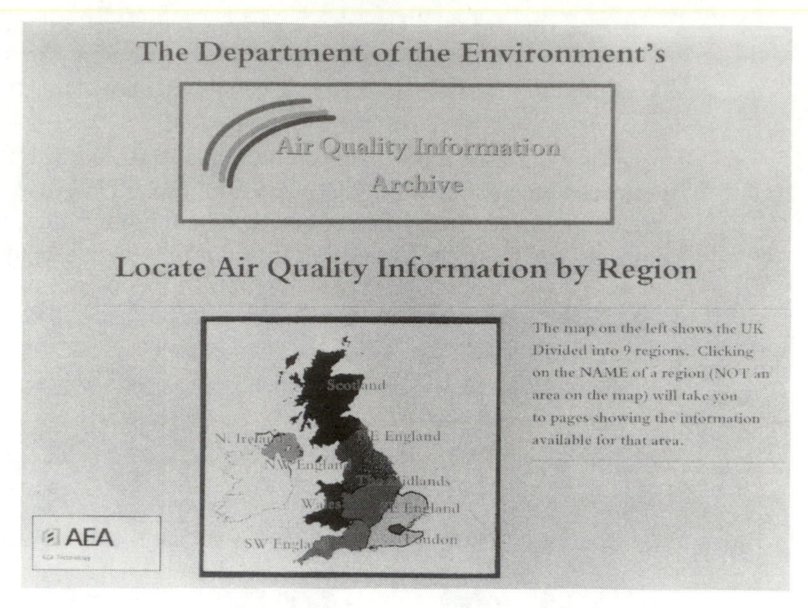


**Figure 13** The world wide web offers a state-of-the-art medium for global data dissemination

now available on the Internet.[9] In the most recent development, the UK's entire archive of air quality and emission data has been made globally available, in a user-friendly and easily downloadable format, on a www site.[10] For the present, this may be regarded as the state-of-the-art in air quality data dissemination.

## 9 A Final Thought

There is a clear need for reliable information on the quality of the air we breathe. This information provides a sound basis for decision-making and air quality management. To be effective, however, measurements must be clearly defined and of documented quality.

The measurement process may be viewed as a chain of activities, as broadly summarized in this review. Quality assurance and control (QA/QC) is an essential feature of any successful monitoring programme. Quality assurance addresses process-related functions such as network design and site selection, management and training systems, and equipment selection. Quality control ensures the integrity of output-related measurement activities; these include site operations, maintenance and calibration, data management, analysis and dissemination. The successful implementation of each component of the measurement chain is necessary to ensure the success of the overall programme (Figure 14).

By reviewing good practice for each operational component of an air survey, this review is intended to assist in obtaining the best possible value—in terms of high quality and useful data—from your monitoring system.

[9] http://www.open.gov.uk/doe/doehome.htm or http://www.aeat.co.uk/netcen/airqual/welcome.html.
[10] http://www.aeat.co.uk/netcen/aqarchive/archome.html.

**Figure 14** Air quality measurement: a final thought

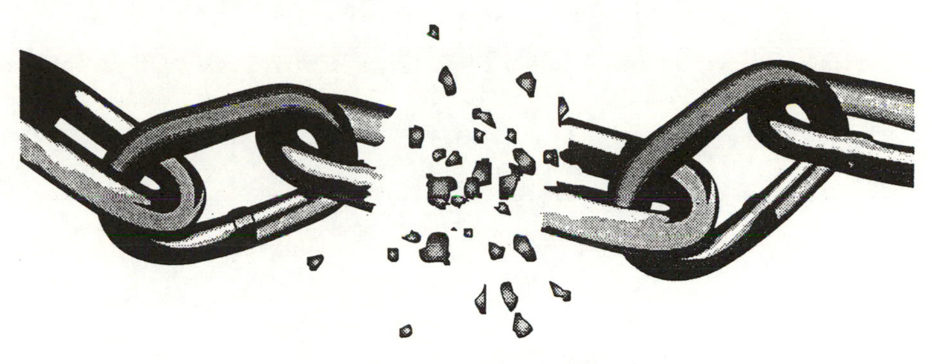

**A Chain is Only as Strong as its Weakest Link!**

## 10 Acknowledgements

This review draws heavily from material prepared by the author for a series of methodology handbooks on air monitoring and QA/QC, published by UNEP's Global Environment Monitoring System (GEMS) and available worldwide. Use is also made of material from NETCEN's operational and quality assurance handbooks for the UK Automatic Urban Network and our air monitoring handbook for UK Local Authorities. We have also drawn freely from our reports on network design for DGXI of the European Commission and on network quality assurance for a recent European Workshop held in ISPRA, Italy, in 1996. The UK's air monitoring networks are funded and supported by the Department of the Environment.

# The European Auto-oil Programme: Scientific Considerations

ANDREAS N. SKOULOUDIS

## 1 The Modelling Framework for Regulatory Decisions

The deterioration of air quality in metropolitan areas is a cause of serious concern in practically all developed nations. Human exposure to high pollutant concentrations has become a worrying issue for national and international authorities. Most of these authorities are currently involved in assessing the status of air quality with the aim to affront 'man-made' pollution by introducing suitable abatement scenarios. These scenarios are frequently linked to reduction in anthropogenic emissions. In the past, new vehicle emission limits have been set using the approach of 'best available technology'. As standards have become increasingly stringent, it has been widely considered necessary to adopt a new approach for the future.

In September 1992 at the Symposium for Auto Emissions 2000 in Brussels, the views of government, industry, environmental groups and others were depicted about the best way to address the problems of air quality. A consensus emerged that future emissions standards should be set within the context of the need to meet environmental objectives, and that consideration should be given to a wide range of measures that can influence emissions from vehicles. This is reflected in the most recent Directives from the European Commission (*e.g.* 94/12/EC), which required further measures against air pollution caused by emissions from motor vehicles which would apply from 2000 onwards. These measures had to be designed to meet the requirements of the Union's air quality criteria and related Directives. An assessment of cost effectiveness was also required, taking into account traffic management, enhanced public transport, new propulsion technologies and the use of alternative fuels.

In order to support the assessment of measures for providing the maximum credibility of foreseen air quality, the Joint Research Centre undertook to utilize the best available technology with the aim to:

- Accurately characterize air quality in seven pilot cities
- Identify real conditions leading to annual mean concentrations and episodes
- Attribute emissions into suitable individual source categories with emphasis on traffic sources

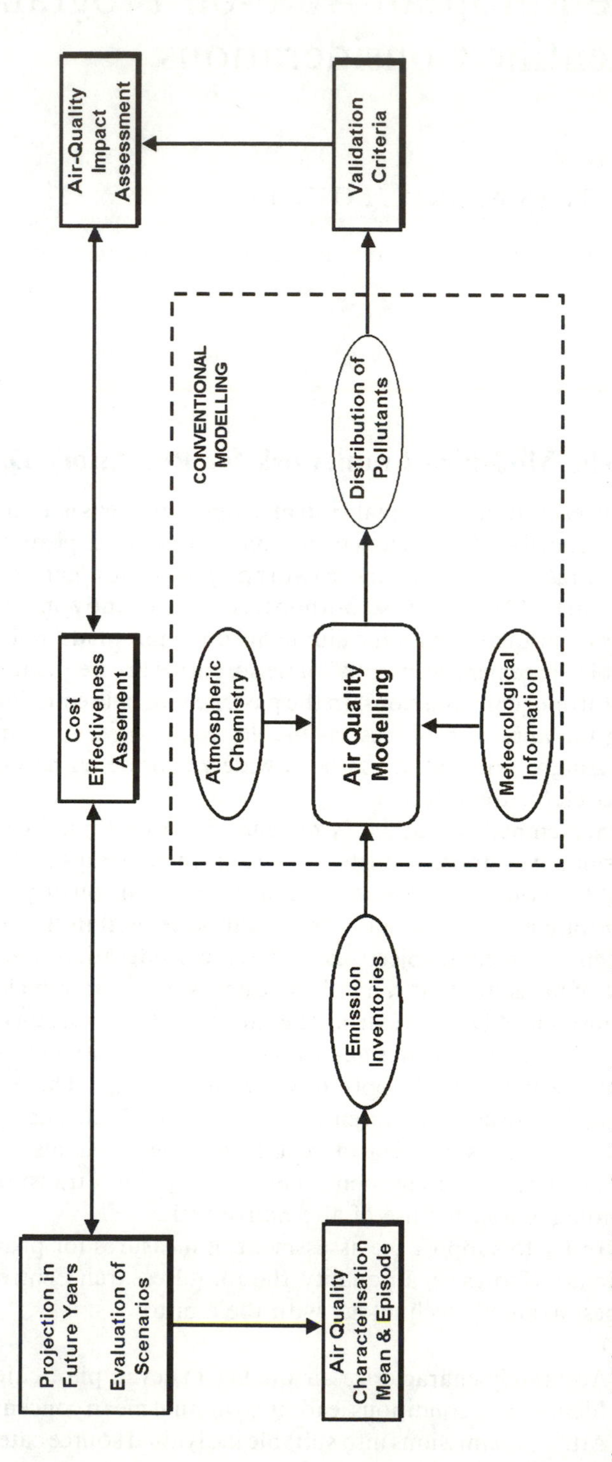

**Figure 1** The methodology for identifying emission reduction targets for each pollutant

- Establish consistent and coherent emissions and topographical databases
- Attribute the concentration of pollutants in the same source categories as in emissions
- Provide a quality assurance of databases and validate the air quality concentrations
- Project the concentrations into the year 2010 when new emission limits will have full impact
- Examine appropriate improvement scenarios based on technological improvements to engines and post-combustion technology, improvements in fuel quality, strengthening inspection and maintenance regimes, *etc.*

In order to achieve the above, a new methodology was required. Schematically this methodology is shown in Figure 1. Six types of models were used in the approach described by this figure. First, the emission forecasting models were used to predict the effect of emissions from already agreed actions and possible future effects of reductions from road transport and stationary sources. This module is indicated by the upper left box in Figure 1. This step was followed by statistical modelling to characterize the air quality conditions over the domain *via* the hourly data collected from monitoring stations. The outcome of this step was a precise identification of the period for which pollution needs to be reduced. From these steps, emissions modelling is needed to describe anthropogenic and natural emissions. The fourth model produces the meteorological data which are then introduced in the dispersion and chemical transformation model. Coupled within this process are the cost-estimate models which are coupled both with the emission forecasting models as well as with the results from the air-quality impact assessment.

Several elements of this approach and the links between them required the construction of modelling modules which are not normally found in conventional models. Another interesting element observed in Figure 1 is the optimization loop between scenarios, air quality and cost assessment. This is the most time consuming process, which is non-linear especially for photochemical pollutants.

## 2 Urban and Regional Models for Air Quality Calculations

For decades, air quality dispersion models have been developed with the aim to simulate and predict the concentration of pollutants in the atmosphere. The main components of these models were:

- Characterization of emissions from natural or anthropogenic sources
- Establishment of consistent meteorological fields over the domain of interest
- Calculation of dispersion of pollutants
- Formation of secondary pollutants (ozone) through chemical transformations

For the models considered it is necessary: to simulate regional and urban scale studies adequately; to account for spatial and temporal characterization of meteorological conditions, turbulence and the planetary boundary layer; to describe dispersion and if necessary the physical transformation (evaporation,

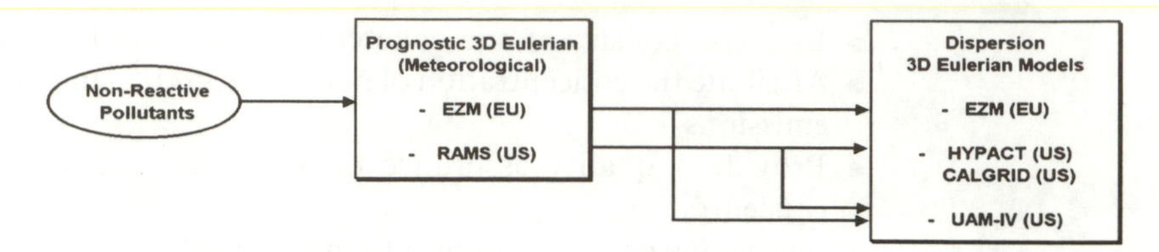


**Figure 2** Schematic representation of the modelling strategy for urban air quality calculations

condensation) of pollutants; to allow chemical transformation of pollutants in gaseous and if possible in aqueous phases (fog or cloud chemistry); to provide good characterization of removal processes (wet and dry deposition) and good surface layer parametrization (soil type, vegetation).

There are several types of models from which to choose. The selection of these models depends primarily on the aims and the requirements of the study to be conducted. For AutoOil-1 the urban scale simulations were conducted with non-reactive pollutants. The aim was to provide reasonable estimates of the concentrations of these pollutants for every hour and in every grid cell during two modelling periods. As described in the following sections, these modelling periods were appropriately selected to represent annual mean and episodic conditions. An important requirement was that the concentrations should be provided in separate source categories with emphasis on traffic sources. With reactive modelling the aim was to examine the effects of emission reductions on ozone concentrations. These concentrations are not normally found in city centres but at some distance away from metropolitan areas. Therefore regional scale modelling was selected for simulating the effects for ozone and its precursors over large parts of Europe.

Based on the results of the urban simulations, the emission reduction targets will be established for achieving air quality standards. Past experience has shown that the main limitation of this approach is usually attributed to the model bias and the user dependency of the results. Since urban simulation was the main process, the strategy followed was to utilize several models which were designed to work consistently. The strategy is demonstrated by Figure 2.

In principle, for the meteorological calculations, two prognostic models were used with the same initial conditions. Prognostic models were used because of the scarcity of surface and upper air measurements, because it was necessary to account as accurately as possible for vertical displacement and because of complex terrain features. As shown by Figure 2, three different dispersion models were used to follow the build-up of pollutants across the city domains. The first two models used the meteorological fields generated by each group, whereas UAM-IV has carried out two separate simulations under the same initial conditions but with the two meteorological fields generated by the two prognostic models.

Description of the main features of the models falls beyond the scope of this work and can be found elsewhere.[1] However, the models used have been of the best available technology with options accounting for:

[1] Subgroup-2 of AutoOil-1, *Air Quality Report of the AutoOil-1 Programme*, European Commission, Brussels, 1996.

- Terrain following co-ordinate surfaces with Cartesian or polar stereographic horizontal co-ordinates
- Cloud micro-physics parametrization at various levels of complexity
- Various turbulence parametrization schemes
- Radiative transfer parametrizations (short and long wave) through clear and cloudy atmospheres
- Two-way interactive nesting with any number of either telescoping or parallel fine nest grids
- Various options for upper and lateral boundary conditions and for finite operators
- Various levels of complexity for surface-layer parametrization (soil model, vegetation, *etc.*)
- Horizontally homogenous or variable initialization (isentropic analysis). Outputs from other models could be used for initialization (*e.g.* ECMWF)
- It is highly portable and runs on several type of computers

For regional scale photochemical modelling a Lagrangian-type of model was used. For this type of model, turbulent dispersion is modelled by simulating the release of a significant number of particles which are advected in response to the mean flow and the turbulence fluctuations. Concentrations are estimated based on the spatial distribution of the released particles with the aid of a univariate or multivariate particle-position probability density function. The EMEP model was selected because it was developed within the UNECE convention on Long Range Transboundary Air Pollution.[2] This is a single-layer trajectory model which calculates regional background concentrations (and other photochemical oxidants) every six hours on a scale of $150 \times 150\,km^2$, covering the whole of Europe. Biogenic emissions of VOCs (treated as isoprene) are calculated every six hours using surface temperature and land-cover data. The advantage of this model was its simplicity for calculating over long time-periods. The EMEP ozone model calculated 15 average concentrations which were assumed to be equivalent to one-hour average concentrations. By assuming that ozone concentrations are highest during afternoon hours, the eight-hour averaged concentrations were estimated by averaging the calculated concentrations at 12.00 and 18.00.

The involvement of several models with different structures and physicochemical options has helped to identify modelling uncertainties and to evaluate the sensitivity of the calculations on the options considered. The scientific co-ordination of the modelling work was entrusted to the Environment Institute of the Joint Research Centre 'Ispra', which is an Institution of the European Commission. The main co-ordinating effort concentrated on providing uniform input data to the modellers. Special attention should be given to protect the calculations from incompatibilities due to different modelling features. Also, for avoiding unnecessary tuning of the input parameters, all calculations should be strictly blind, *i.e.* modellers should not know results from each other before simulations for each test city have finished. For the same reason, the experimental data and the

[2] D. Simpson, *Atmos. Environ.*, 1992, **26A**, 1609.

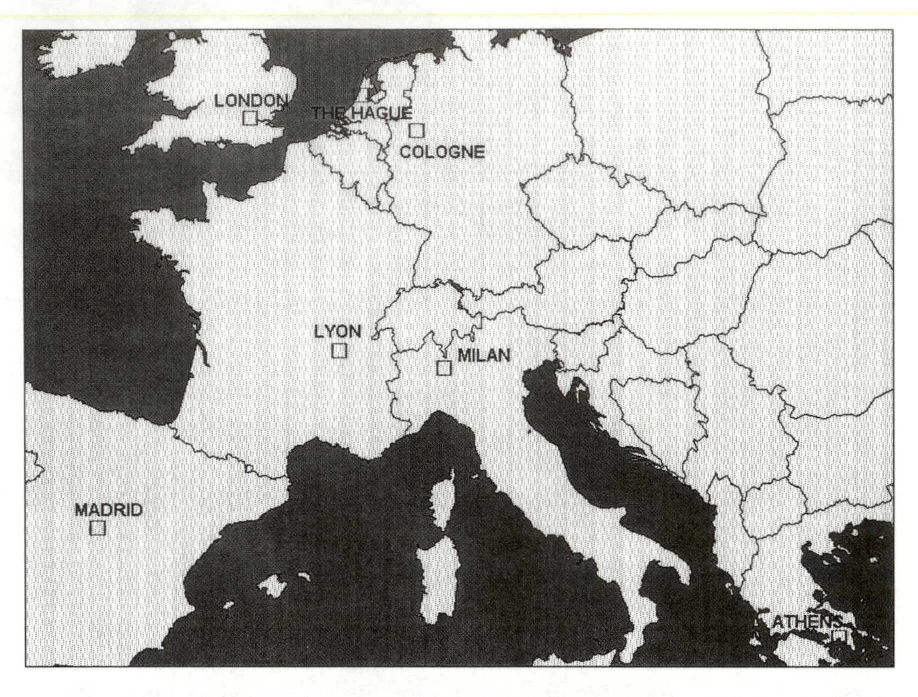


**Figure 3** Geographical details for the seven domains of urban air quality simulations

| City | Country | Population (1990) | Area (km²) |
|---|---|---|---|
| Athens (Attiki) | Greece | 3523000 | 3800 |
| Cologne | Germany | 3963000 | 7400 |
| Hague | Netherlands | 692000 | 200 |
| London | UK | 6794000 | 1600 |
| Lyon (Rhone) | France | 1508000 | 3200 |
| Madrid | Spain | 5028000 | 8000 |
| Milan | Italy | 3987000 | 2800 |

validation tests were not known to the modelling groups prior to the actual simulations.

## 3 Characterization of Air Quality Data from Monitoring Stations

For urban air quality simulations, seven cities were selected which were representative in terms of air quality, location and size. These cities are shown in Figure 3 together with the relevant geographical details. In terms of air quality these cities were representative of a range of large cities found in Europe.[3] The problem which arises now is to find which of the monitoring stations within each city can be used to characterize the overall city air quality. There are several approaches concerning this problem, but in practice there was dis-equilibrium concerning the number and the functionality of the monitoring stations between northern and southern cities. So for reasons of simplicity a monitoring station at

[3] C. Holman, *Current Air Pollution due to Transport Activities (Regulated and Unregulated Exhaust Emissions) in Selected Cities in the European Community*, European Commission, Brussels, 1994.

the city centre was selected to represent each city. Those stations were Athinás (Athens), Riehl (Cologne), Constant Rebecque Straat (The Hague), Victoria Bridge Place/Central London (London), Garibaldi (Lyon), G.T. de Carlos V (Madrid) and Cenision (Milan).

In order to estimate the annual mean concentrations in an urban domain with a grid of a resolution of $2 \times 2\,km^2$, approximately 25 000 cells are required. For these cells the meteorological fields and concentrations need to be calculated on an hourly basis. It is inevitable that the simulation period considered cannot be extended to cover a whole year. Besides, neither the emissions nor the initial meteorological conditions are known with sufficient precision that this calculation will have a significant meaning.

However, owing to the recurring nature of meteorological conditions and the repetitive characteristics of the anthropogenic emissions, we could identify several real periods during which the mean concentration corresponds to annual mean values. The period which adequately describes the annual mean concentration can be determined by calculating the rolling average concentrations for periods of duration ranging from 1 to 21 days. For each pollutant and each rolling period the mean concentrations were determined. These were then compared with the mean concentration for the whole year. Each rolling period was ranked according to its deviation from the annual concentration. These deviations were summed up for each period for both CO and $NO_x$. The period with the smallest deviation from the annual mean was chosen as the surrogate period representing annual mean conditions.

This analysis, carried out for rolling averages of different duration, indicated similar dates in 1990 when the concentrations were averaged for longer than four days. The final selection and the corresponding mean values are shown in Table 1. The process for the city of The Hague is based only on the data of $NO_x$ since CO was not measured in 1990. The same process was repeated for 1988 for The Hague, which was the last year when CO was measured. The concentrations for the surrogate mean and the actual annual mean values are respectively 40.108 *versus* $40.111\,\mu g\,m^{-3}$ for $NO_x$ and 434.55 *versus* $434.11\,mg\,m^{-3}$ for CO. The concentrations for $NO_x$ are comparable with those of 1990. The last two columns of Table 1 show the percentage of valid data during the base year for the monitoring station selected at the city centre.

**Table 1** Comparison of annual mean concentrations with the mean values for the surrogate period for the modelling calculations

| | | *Surrogate mean* | | *Annual mean* | | *Valid data* | |
|---|---|---|---|---|---|---|---|
| *City* | *Initial date* | CO/ $mg\,m^{-3}$ | $NO_x$/ $\mu g\,m^{-3}$ | CO/ $mg\,m^{-3}$ | $NO_x$/ $\mu g\,m^{-3}$ | CO/ % | $NO_x$/ % |
| Athens | 30 Mar | 4.1 | 160.1 | 4.1 | 159.7 | 100 | 100 |
| Cologne | 18 Jul | 1.1 | 94.9 | 1.1 | 94.6 | 99 | 96 |
| The Hague | 26 Jun | — | 40.7 | — | 40.7 | — | 100 |
| London | 13 Jan | 1.1 | 77.8 | 1.1 | 77.9 | 98 | 99 |
| Lyon | 25 Feb | 4.6 | 175.8 | 4.6 | 197.8 | 86 | 100 |
| Madrid | 05 Mar | 2.2 | 215.0 | 2.2 | 214.2 | 93 | 95 |
| Milan | 15 Feb | 3.3 | 325.5 | 3.3 | 318.5 | 94 | 100 |

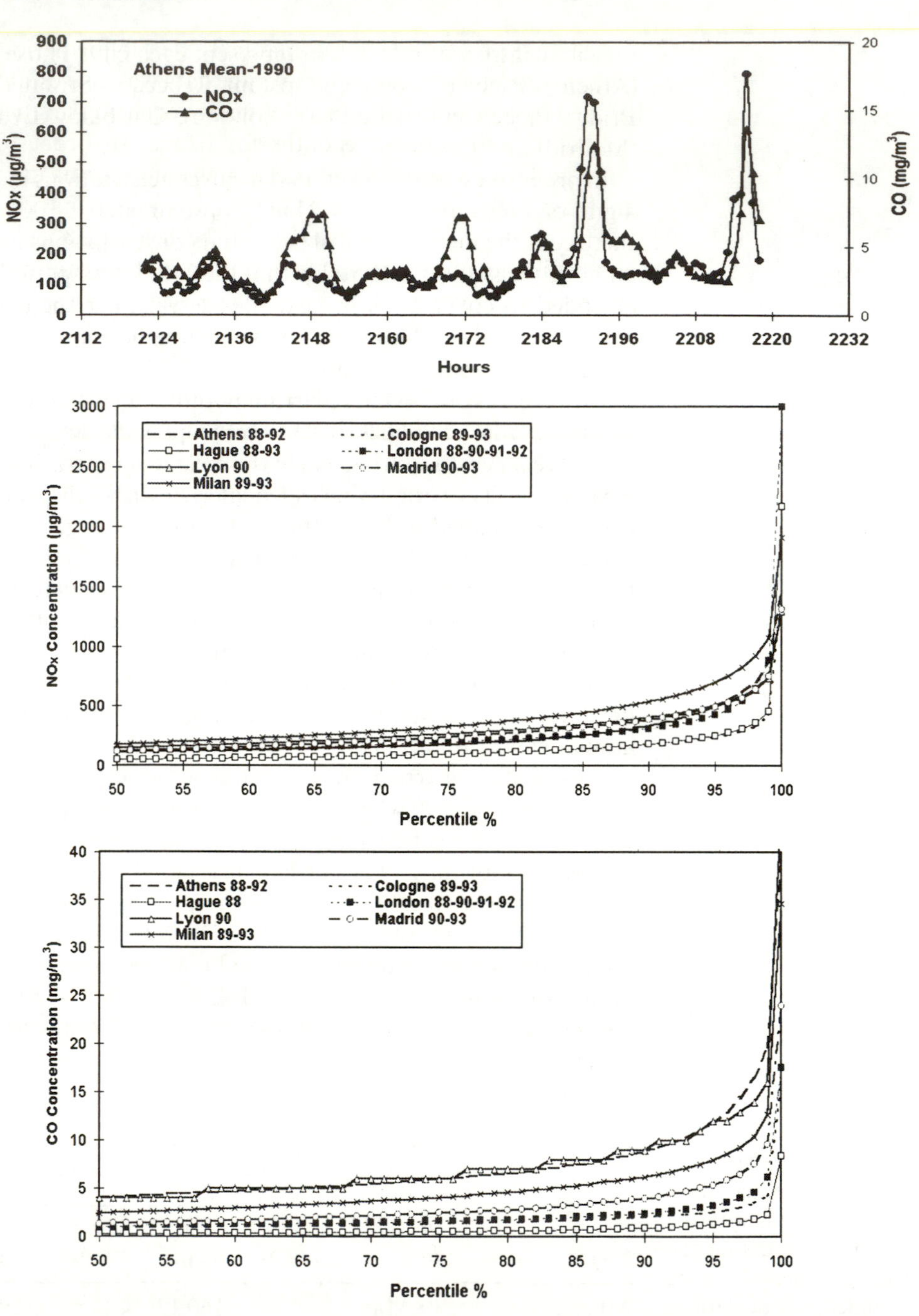


**Figure 4** Concentrations at the Athinás monitoring station during the surrogate annual mean period

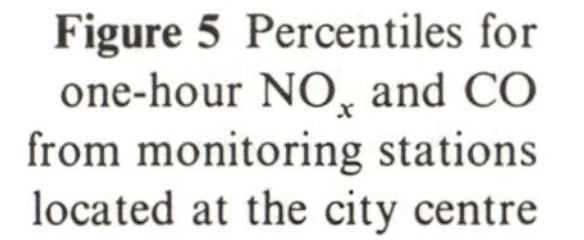
**Figure 5** Percentiles for one-hour $NO_x$ and CO from monitoring stations located at the city centre

The concentrations shown in Figure 4 are the actual concentrations recorded in Athens during the period depicted by the above procedure as representative of annual mean conditions. The advantage of this approach is that there is no need to define artificial diurnal variations and the emission inventories could be tailored to represent actual anthropogenic activities for the period under consideration.

A similar procedure for automatically identifying periods of elevated pollution concentrations from one-hour average data was also established. The periods are defined as 'episodes' and can frequently occur during winter months for inert pollutants and summer months for ozone. In order to identify episodes of different severity and duration accurately, monitoring data for several years were analysed.

The severity of the episodes can be selected according to the percentile characteristics of each monitoring station. The percentiles for $NO_x$ and CO of the seven cities are shown in Figure 5. Based on this figure, inter-comparisons between the seven cities should be avoided owing to different representativeness of these stations at each city. This figure shows the change of concentrations above the 90 percentile. Based on these plots it is clear that elevated concentrations cannot only be recognized from the values recommended by the international air quality standards. For some cities like The Hague these standards are rarely exceeded, whereas periods exceeding the 90 percentile show conditions where deterioration of current air quality is evident.

With this approach it is also possible to identify the length of the episode. This was defined by the number of hours above the concentration corresponding to the selected percentile. Episodes are considered to be distinct when there are more than 10 hours between them with concentrations below the selected limit value.

For each distinct episode the maximum one-hour concentration is defined for each pollutant. The episodes were ranked for $NO_x$ and CO separately with the highest concentration having the lowest ranking. By adding the ranks for each pollutant, episodes of different severity and duration can be identified according to the lowest combined ranking. *A priori* to the selection it might be necessary to exclude episodes of 'extreme nature' which have a very low probability to occur again.

Details about this process can be found.[4] This characterization process can be carried out for primary or secondary pollutants. By analysing episodes of different duration it was shown that episodes of short duration are characterized by intermediate concentrations with a high frequency of occurrence. On the other hand, during long episodes, peaks of high concentrations are expected in addition to more frequent intermediate values. Another aspect of this work is that this process can be carried out for one or many monitoring stations. By repeating this process in a domain with 30 monitoring stations and after the exclusion of non-operational stations, all other stations consistently indicated the same dates for episodes of equal severity.[5]

## 4 Construction and Disaggregation of Emission Inventories

Issues related to emission databases are mainly logistic in nature. It is necessary to cover a large region for the urban modelling domain and the whole of Europe

[4] R. Bellasio and A. N. Skouloudis, *An Identification Methodology for Worst Pollution Episodes in Urban Atmospheres*, Joint Research Centre Ispra, European Commission, Technical Note I.97, Ispra, 1997.

[5] A. N. Skouloudis, R. Bianconi and R. Bellasio, *Environ. Monitoring Assessment*, 1997, in press.

for the photochemical modelling. The methodology adopted was to use data from relevant responsible local, regional or national organizations. The disadvantage is that emissions are attributed to sources which are not compatible and in most cases not sufficiently disaggregated to address all traffic sources necessary for AutoOil-1. These categories were: passenger cars—gasoline (PCg), passenger cars—diesel (PCd), light duty commercial vehicles (LDV), medium and heavy duty commercial vehicles (MHDV), large stationary sources (LSS), medium stationary sources (MSS) and small stationary sources (SSS).

For the city emission inventories the modelling module shown in Figure 1 includes features for

- Spatial disaggregation for modelling domains of 10 000 $km^2$ with a resolution of $2 \times 2 km^2$
- Temporal disaggregation of emissions for each hour of weekdays and adjusted to reflect monthly variations
- Source harmonization of emission rates for the above mentioned source categories of AutoOil-1 in each of the grid cells
- Validation and quality assurance of the emission databases constructed

The area covered by the urban domains for all cities examined are larger than the area covered by the city itself. Hence, it was necessary to account for background emissions or other urban centres included in the domain as, for example, for The Hague. Apart from the original inventories requested from relevant national organizations, the emission data from CORINAIR-90 were consulted at the highest possible grid or administrative area resolution.

Usually emission data are given on an annual basis. Temporal disaggregation should be carried out with the aim to determine, at first, appropriate daily emissions. This is an important step of the overall modelling process since air quality simulations were conducted under real conditions on specific dates in 1990. Diurnal variations were taken into consideration according to each individual source category. The variations for traffic sources taken into consideration are traffic density, fleet composition (type of vehicles, size of engine, age, *etc.*) and traffic speed. For other sources the national operating characteristics were assumed. For example, large stationary sources were assumed to operate on a 24 hour basis and the medium stationary sources for 16 hours per day. A typical example of the diurnal multiplication factors is shown in Figure 6 for $NO_x$ in the city of Milan and the sum of each of these multipliers over 24 hours is one.

Similar multiplication factors and explanations on where these were based for each city can be found at the Report of Subgroup 2.[1] In the same report, detailed explanations are also given about the source attributions and how these were achieved for each grid cell. The following figures depict, for each city, the attribution of emissions for the first 50 cells with the highest emissions. The area covered by these cells is 200 $km^2$, which represents 2% of the modelling domain. The number at the $x$-axis of Figures 7 and 8 represents the rank number and not the exact position over the domain.

The emissions shown in Figures 7 and 8 correspond to the surrogate annual mean period identified by the previous sections. For this reason the emissions from small area sources are absent during the summer air quality simulations.

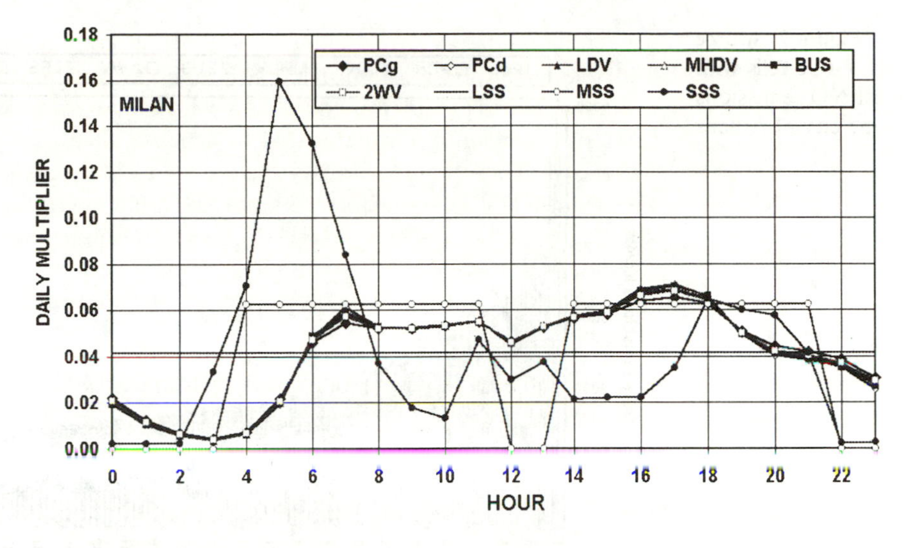


**Figure 6** Multiplication factors for the diurnal variation of emissions over the whole domain

The contribution from large stationary sources is also dominant for some grid cells and absent from the domain if large industrial sources are not present.

Until now, the shortcomings from air quality calculations have been attributed to inaccurate emission databases. Emission data, derived from pure logistic processes, can rarely be validated with measurements. This is mainly due to the fact that soon after the release of a pollutant, dispersion, chemical transformation and other interactions with urban processes take place. For this reason the validation of emissions is a desirable process. The validation of the emission databases can be carried out by comparing databases constructed with different methodologies or with emissions aggregated either in time or in space.

The parameter for validating the emission inventories utilized is the emission flux, that is the emission mass emitted per unit time and unit area. Taking into consideration that most of the emission inventories are generated by different projection systems and refer to different grid cell resolutions, the validation of emissions is a laborious process. For the purposes of AutoOil-1, comparisons have been carried out with the annual data of CORINAIR-90. This process for most of the cities used for urban modelling showed differences up to 20% for city centres. Similar comparisons can be carried out with emission inventories constructed with an alternative approach, for example with a 'bottom-up' aggregation. In the latter case the emissions inside each individual cell are considered separate polluting sources, *e.g.* for each road or each industry, which are then aggregated over the whole domain.

The emission data for several pollutants were needed for conducting air quality simulations for ozone. The emission data were taken directly from the EMEP databases.[6] The emission source categories and the methodology are identical to those used for CORINAIR-90.[7] The emission data were gridded by the European

[6] E. Berge, H. Styve and D. Simpson, *Status of the Emission Data at MSC-W*, EMEP MSC-W, Report 2/95, Norwegian Meteorological Institute, Meteorological Synthesizing Centre, Oslo, 1995.

[7] *Reference Guidebook on Emission Inventories*, UNECE/CORINAIR, European Environment Agency, Copenhagen, 1995.

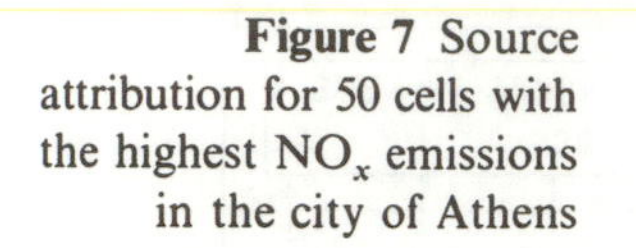

**Figure 7** Source attribution for 50 cells with the highest $NO_x$ emissions in the city of Athens

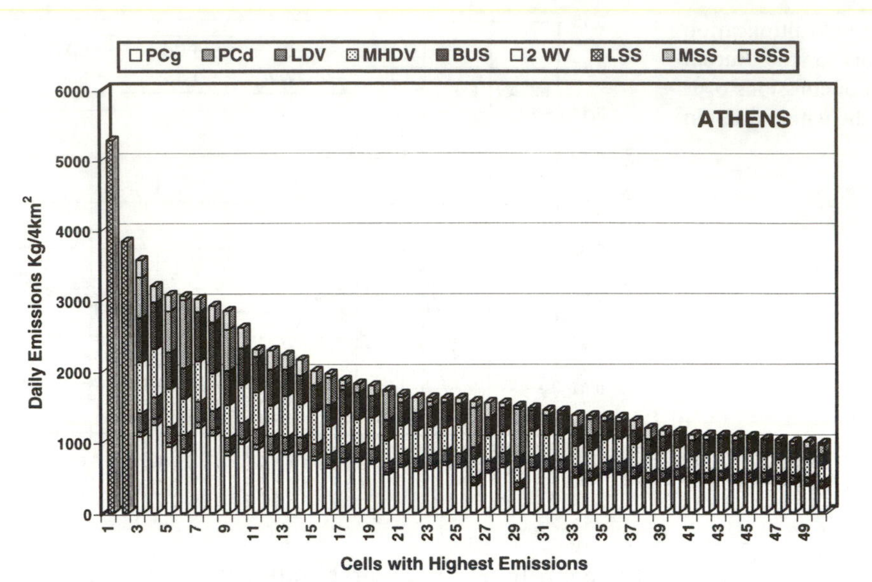


**Figure 8** Source attribution for 50 cells with the highest $NO_x$ emissions in the city of Lyon

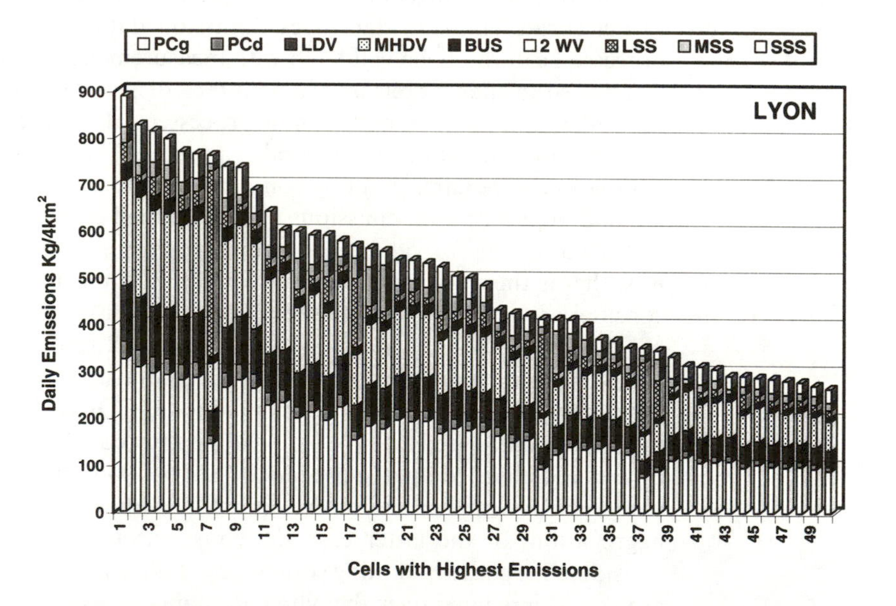


Environment Agency at a database of 50 × 50 km$^2$. The total emissions for EU12 countries are shown in Figure 9. Concerning biogenic emissions of VOCs, these were calculated on-line at the EMEP model using isoprene as a surrogate for all species. The hourly calculation of isoprene was carried out using the model's radiation model, the local temperature and the land-use data. In many cases these were different from CORINAIR-90 and this is an area where further work is needed for future validation. This area of uncertainty was examined by a sensitivity analysis carried out under the same modelling conditions.

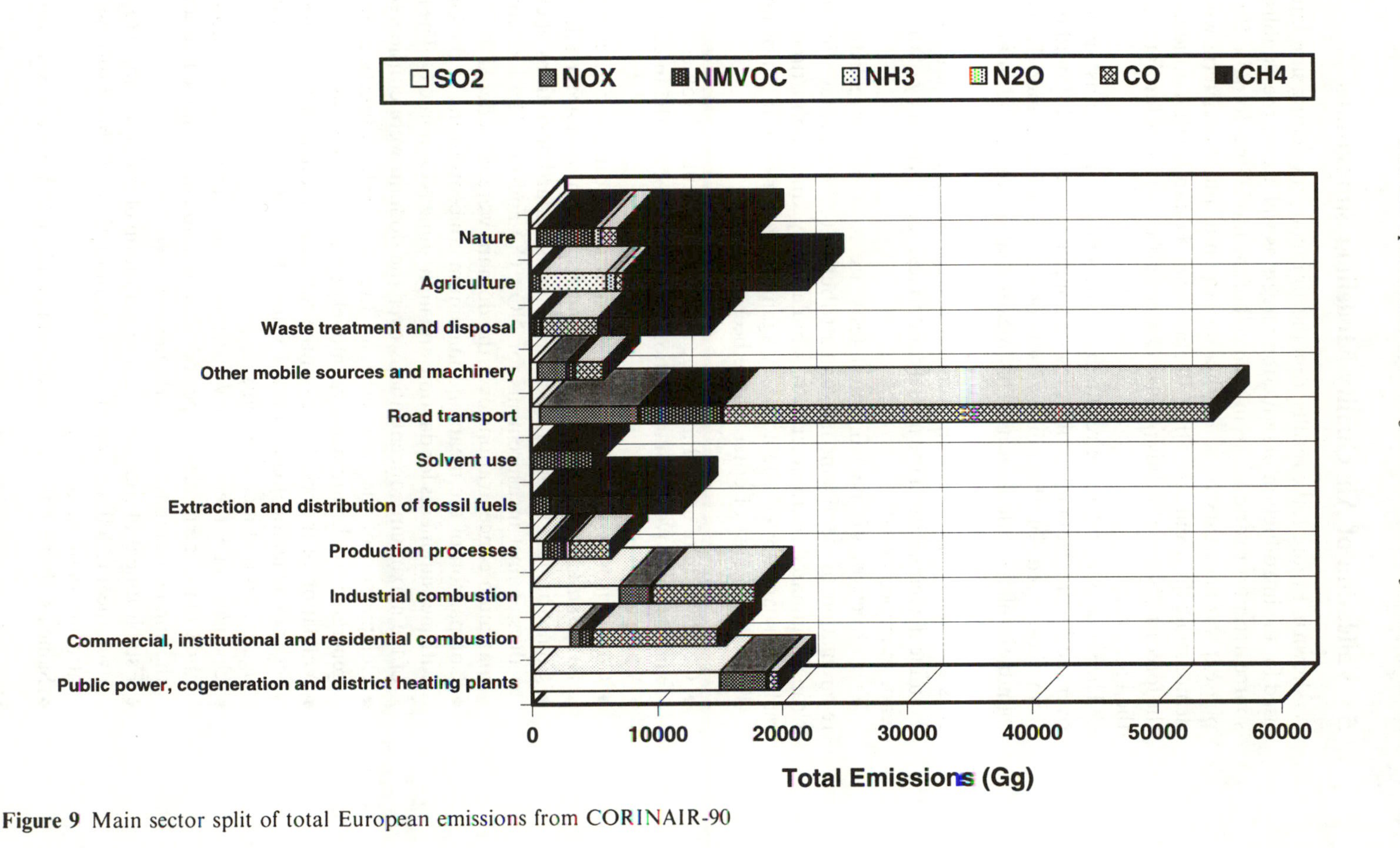


**Figure 9** Main sector split of total European emissions from CORINAIR-90

## 5 Validation of Air Quality Modelling Simulations

As shown in Figure 1, the results from the air quality modelling simulations were used for the introduction of reduction targets as well as for the establishment of source–receptor relationships in each cell of the modelling domain. Prior to this phase it was necessary to establish a set of non-disputable criteria which could demonstrate the accuracy of the calculations and establish the confidence level of the forecasts. Naturally, this process can be carried out only for data referring to the base year.

The scope is to compare results from the simulations against data collected over the domain from all monitoring stations. The data used should be different from those used for the initialization of the air quality models. Hence the 'blindness' of the calculations. This meant that, prior to the final evaluation, comparisons and adjustment of modelling results to measurements were avoided. It is also desirable to establish confidence levels against monitoring data, to examine the reliability of the models and to test the consistency of the modelling features implemented.

The objectives of the validation tests are to avoid unrealistic modelling re-circulation, to identify inconsistent attribution to source categories, to identify wrongly initiated parameters, to simulate day and night time differences realistically and to test the accuracy against time series of monitoring data.

The following list of criteria have been used:

- Comparison of annual mean measured and calculated values of $NO_x$, CO and benzene over as many stations as possible or for the integral over the whole city
- Comparison with measurements of diurnal air quality concentrations for $NO_x$ and CO for monitoring stations for which data are available
- Comparison of the mean, maximum and standard deviations of calculated values against measurements for $NO_x$, CO and $O_3$. The locations of the maximum concentrations over the domain were considered
- Comparison of the mean and maximum concentrations at the city centre with measurements for daytime and night-time $NO_x$ and CO. The locations of the maximum concentrations over the domain were considered
- Comparison of the time series of the wind fields (direction and speed) generated by the atmospheric models on an hourly basis
- Validation of the wind fields of the surrogate annual mean period against measured meteorological data. Examination of the frequency of occurrence against annual measurements
- Comparison of the frequency distribution of hourly concentrations of $NO_x$ and CO for the modelling period. Examination of the same frequency from the annual database of hourly measurements
- Establishment of source category attribution of emissions for $NO_x$ and CO over the modelling domain and examination of this ratio against the attribution of air quality
- Same as above but separate daytime and night-time concentrations

Naturally, if the number of the above criteria is multiplied by the number of

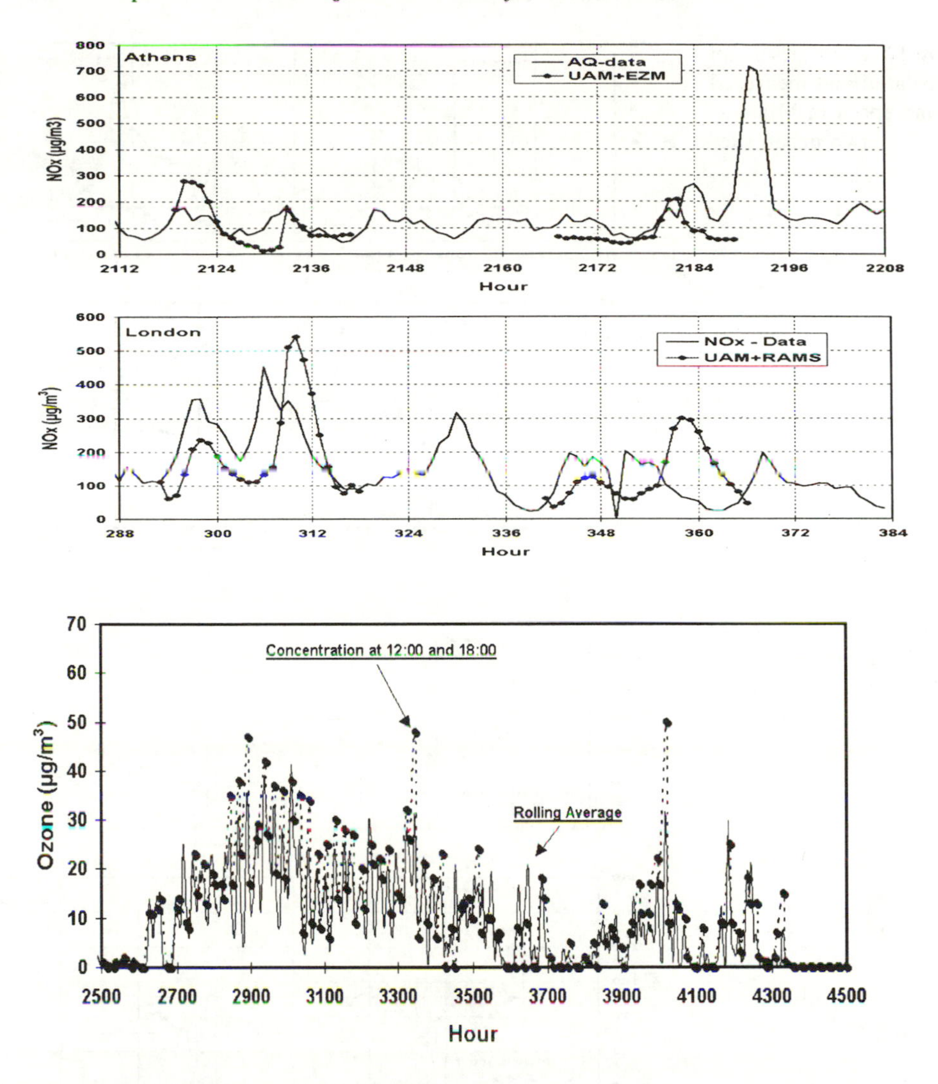


**Figure 10** Time series of the $NO_x$ concentration at the monitoring station of Athinás (Athens) and Bridge Place (London). The hours indicated are from the start of 1990

**Figure 11** Comparisons of ozone at 12.00 and 18.00 with rolling eight-hour average from monitoring data at Station Garibaldi in Lyon (1990). The hours indicated are from the start of the year

pollutants, the number of models and the number of cities, the inter-comparisons which could be presented are more than 2100. For purely logistic reasons it is only possible to show here a limited number of randomly selected results. These are shown in Figures 10–14 starting from the time series of concentrations for $NO_x$ and ozone.

Figures 12 and 13 show comparisons from the atmospheric models for two urban monitoring sites in Milan. Figure 14 is an overview of the frequency distributions for $NO_x$ in all cities. In the latter set of figures are shown the actual measured data during the surrogate annual mean period, the corresponding calculated concentrations from the best model and the histogram of $NO_x$ concentrations obtained from the annual data for the base year. Comparisons between the annual data and the data for the surrogate annual period demonstrate how representative is the period selected for annual mean conditions. Comparisons between the measured and calculated data show the accuracy of the

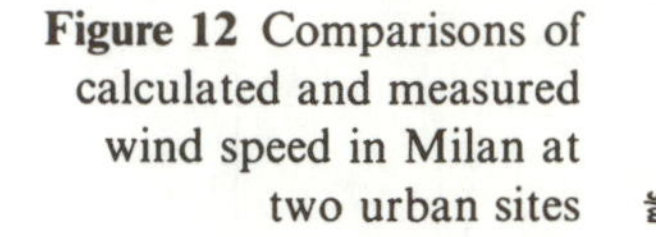

**Figure 12** Comparisons of calculated and measured wind speed in Milan at two urban sites

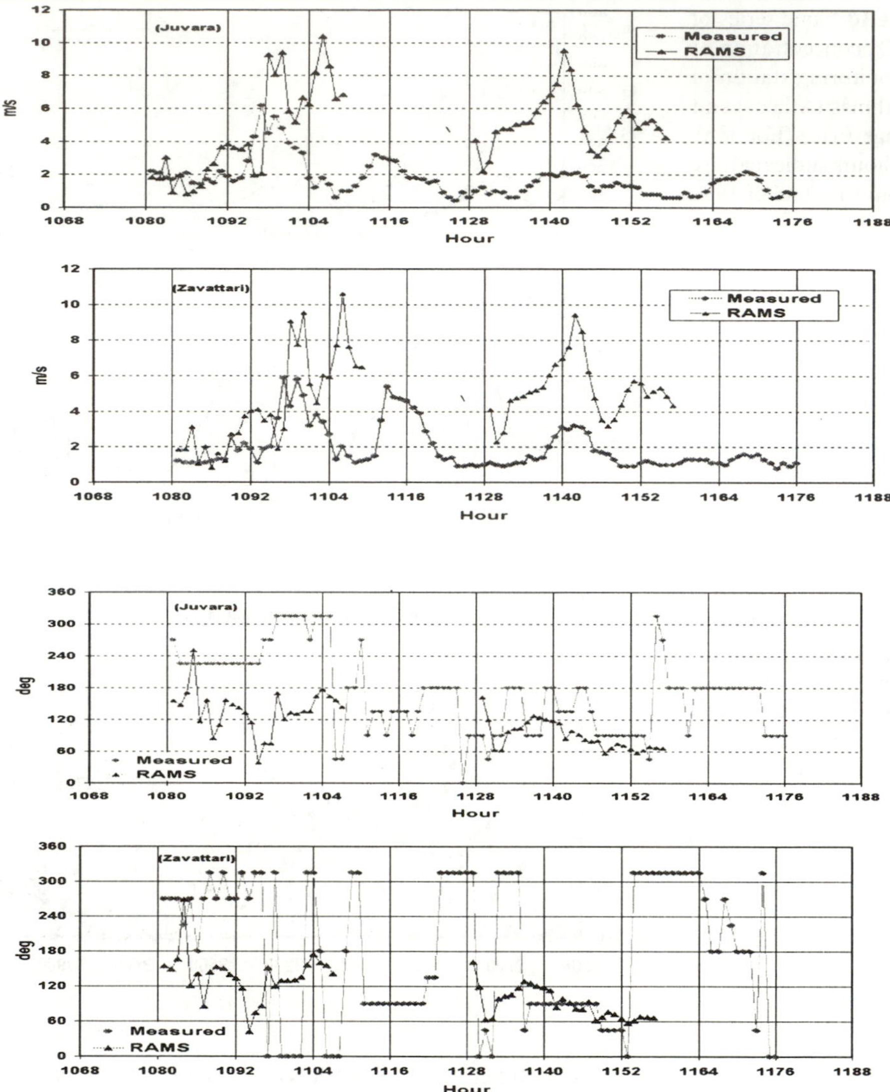


**Figure 13** Comparisons of calculated and measured wind direction in Milan at two urban sites

simulation. Naturally, these plots show only the temporal compliance. For the spatial compliance the same comparisons need to be carried out for all monitoring stations in the calculating domain.

In addition to the above figures, and whenever possible, the parameter of interest in the validation criteria was tested against the ratio of '(predicted – observed)/observed' values. A typical comparison of this type is shown in Figure 15. From these plots the simulations with most points within the 50% lines can be readily selected. Figure 15 shows comparisons for the same city but with two different models, with the second model over-predicting city centre concentrations and the source attribution during night-time.

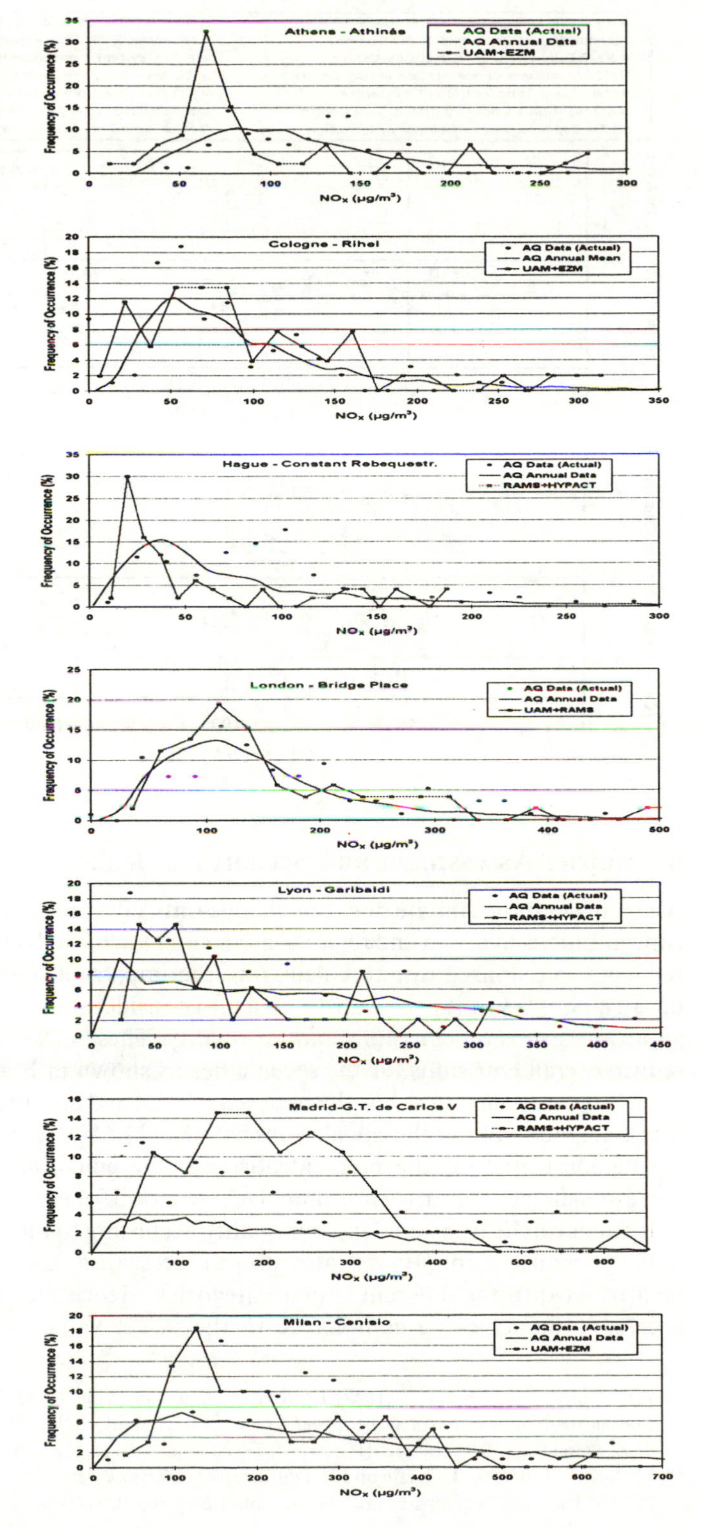


**Figure 14** Histograms of measured and calculated $NO_x$ concentrations during the surrogate annual mean period for each city. The histogram of the annual concentrations is plotted for comparisons

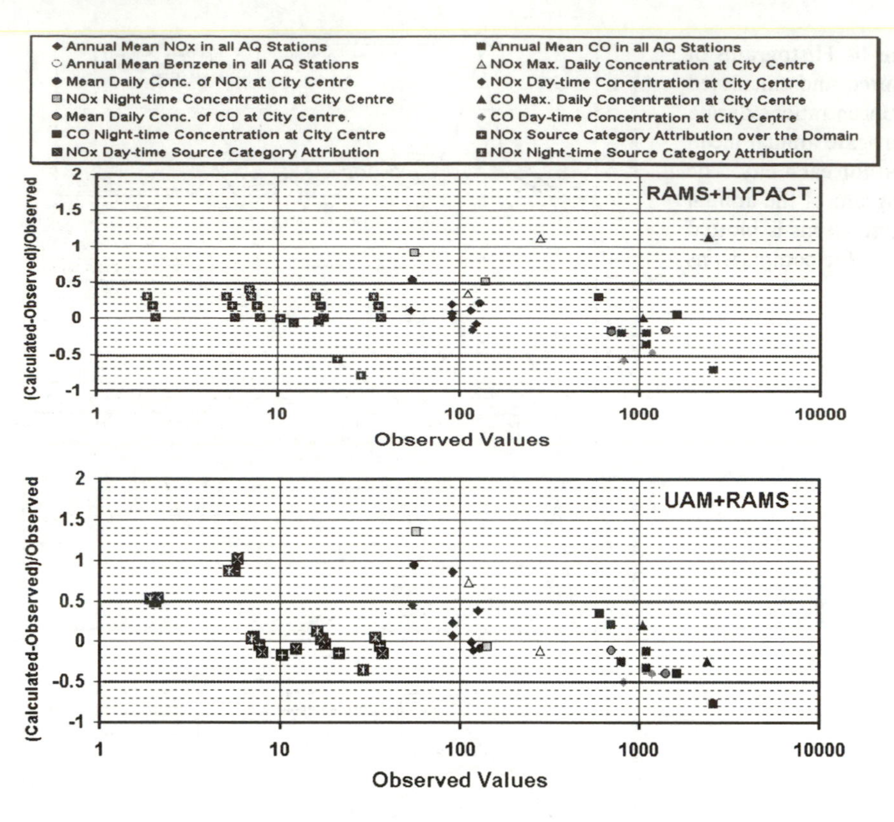


**Figure 15** Comparisons of calculated and observed parameters from two different sets of models for the city of Cologne

## 6 Impact Assessment and Scenario Calculations

Assuming that the climate does not change until 2010, the reduction of pollutant concentrations in urban and regional areas can be achieved only by appropriately reducing emissions. Prior to calculating the impact of reduction scenarios in urban air quality, it is necessary to take into consideration the impact of already agreed measures on emissions and air quality. These measures are expected to reduce overall emissions for the seven cities as shown in Figure 16. This figure depicts both the emissions for the base year (1990) and also the forecasts for 2010. These are the results of the calculations with FOREMOVE[8,9] and RAINS[10] and are demonstrated for the nine categories of the emission inventory for $NO_x$ corresponding to the surrogate annual mean period for each city. These emission reductions can be translated into air quality according to the approach described in the previous section. By repeating this process, air quality concentrations can be attributed to the different source categories. Hence the amount of emission necessary for achieving compliance to the target values set by international

[8] Z.C. Samaras and K.-H. Zierock, *Forecast of Emissions from Road Traffic in the European Communities*, Commission of the European Communities Report, EUR 13854, DG-XI, Brussels, 1992.

[9] T. Zachariadis, Z. Samaras, K.-H. Zierock, *Technolog. Forecasting Social Change*, 1995, **50**, 135.

[10] M. Amann, I. Bertok, J. Cofala and P. Dorfner, *$NO_x$ Emission Attenuation Profiles up to the Year 2010 in Europe*, The Trans-boundary Air Pollution Project, IIASA, Laxenburg, 1990.

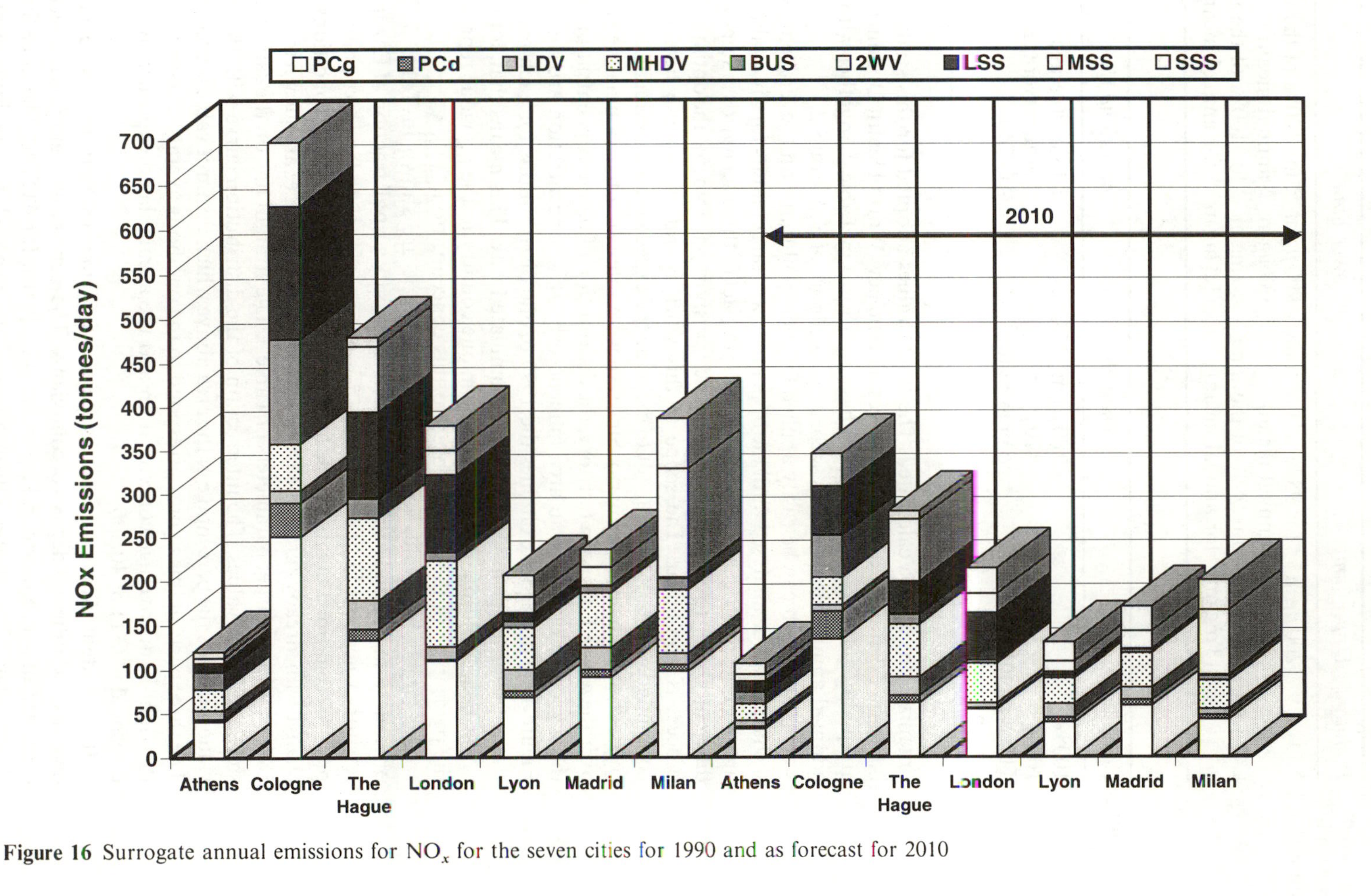


**Figure 16** Surrogate annual emissions for $NO_x$ for the seven cities for 1990 and as forecast for 2010

**Table 2** Upper and lower limit values utilized in AutoOil-1

| *Pollutant* | *Upper limit* | *Lower limit* |
|---|---|---|
| $NO_2$ | 200 μg m$^{-3}$ for 98% of the time (78 μg m$^{-3}$ annual mean) | 200 μg m$^{-3}$ for 100% of the time (37 μg m$^{-3}$ annual mean) |
| CO | 10 mg m$^{-3}$ for 98% of the time (3 mg m$^{-3}$ for annual mean) | 10 mg m$^{-3}$ for 100% of the time (1.5 mg m$^{-3}$ for annual mean) |

**Table 3** Target values for ozone

| *Compliance for* | *Ozone value* | *Averaging time* |
|---|---|---|
| 100% | 180 μg m$^{-3}$ | 1 hour average |
| 100% | 120 μg m$^{-3}$ | 8 hours average |
| Three months growing season | 5 ppm hours | AOT40 |

organizations can be calculated. The target values adopted for this study were based on existing air quality objectives and the new World Health Organization air quality guidelines for Europe. These are shown in Table 2, together with the adopted annual mean values according to report of Subgroup 2.[1]

This approach can be easily applied for inert pollutants and several emission scenarios can be analysed until the upper or lower target limits for air quality are met. For the photochemical pollutants the impact of these scenarios is somewhat different. Owing to chemical reactions and transport phenomena, the optimization process is non-linear. The scope of this work is not to analyse how ozone compliance can be achieved, but to examine the expected improvement in ozone concentration by each scenario. Since we were interested to examine the effects over the whole European territory, the distribution of ozone concentrations and the percentile of European area influenced by this concentration are shown in Figure 17. This figure examines the 1990 ozone distributions for each member state of the European Union in an accumulated way. This means that, for each ozone concentration, the difference from one country to the previous represents the percentage of the area exposed to this concentration.

The targets for evaluating ozone over Europe are shown in Table 3. It is obvious from Figure 17 that calculations show only a small fraction of European territory where the limit of 180 μg m$^{-3}$ is not exceeded. As for inert pollutants, prior to scenario calculations, the first objective is to see how these conditions will change with already agreed measures by 2010. These changes are shown in Figure 18 for the three target parameters of Table 3. The straight line on the same figure shows the limit values. The intersection of the distribution curve with the limit value indicates the percentage of the area above which the limit value is exceeded. The three parts of this figure correspond to the one-hour average concentration, the eight-hours average concentration and the accumulated exposure over the threshold (AOT) of 40 ppb.

It is evident from the various parts of Figure 18 that the three limits demonstrate different degrees of compliance. The eight-hour average is practically exceeded all over Europe and the same is nearly true for AOT40. The same figure shows that 73% of the European territory exposed to one-hour average

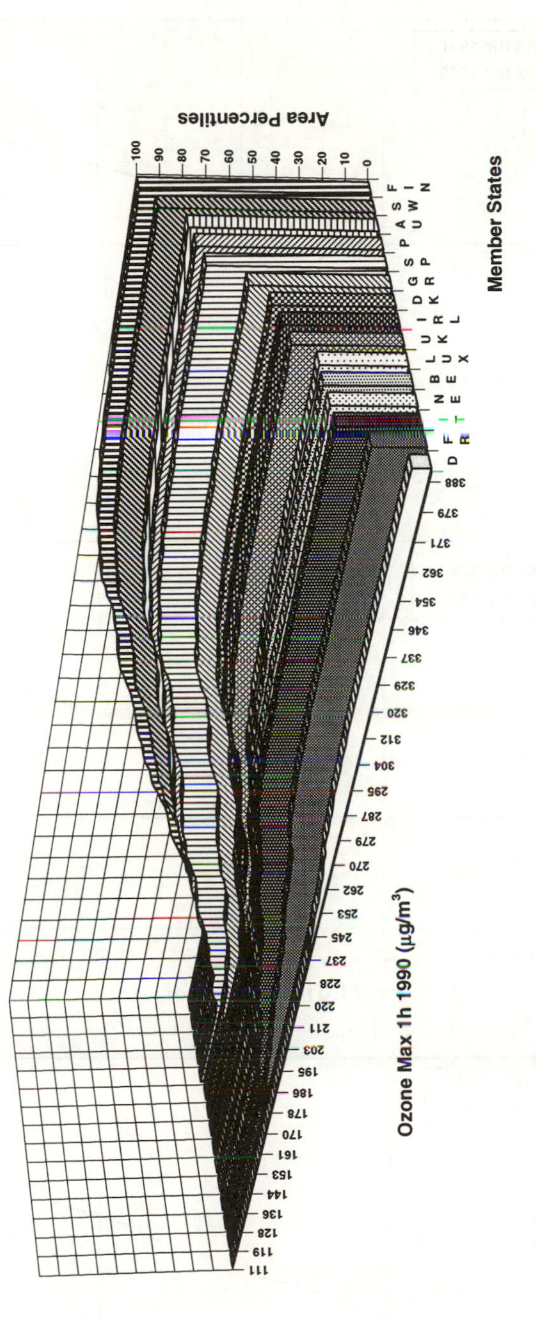


**Figure 17** Maximum one-hour ozone concentrations and the percentage of land cover affected by this concentration at the 15 member states of the European Union

**Figure 18** Ozone concentrations in 1990 for the 15 EU states and the percentage of land cover affected together with the impact of agreed measures until 2010 on the maximum one-hour, maximum eight-hour and AOT40 concentrations

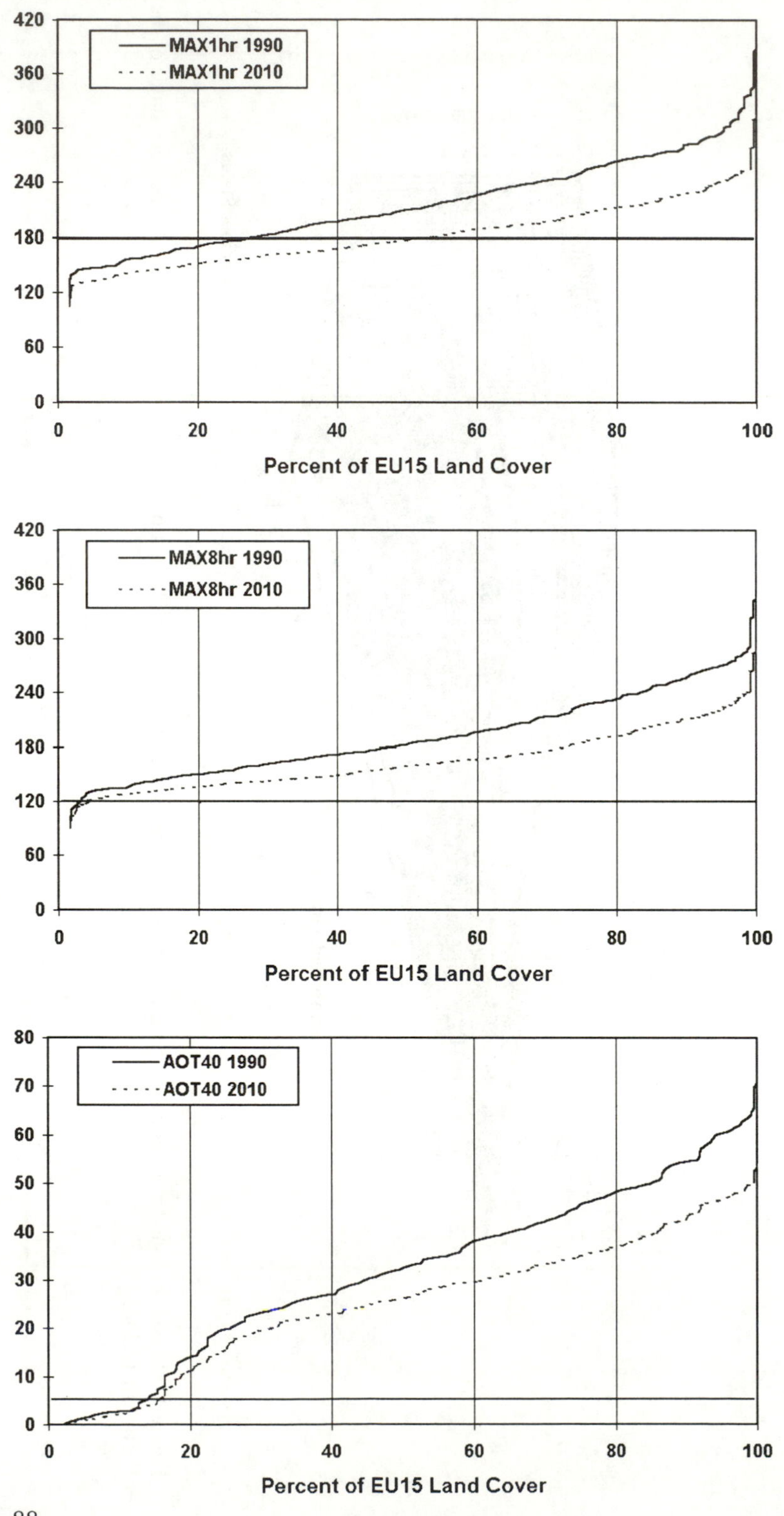

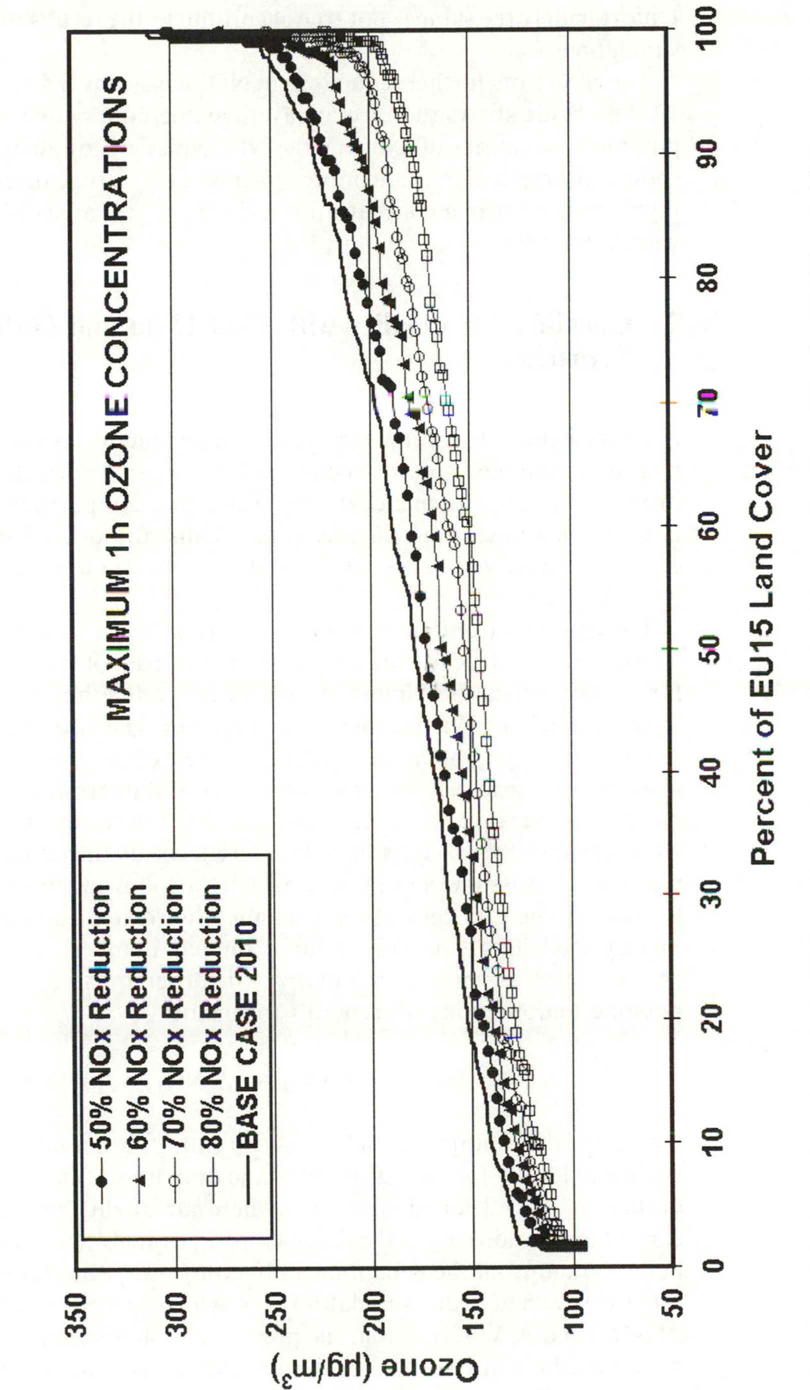

**Figure 19** Effect of further $NO_x$ reductions in 2010 on the maximum one-hour ozone concentrations over the 15 EU countries

concentrations of 180 $\mu g\,m^{-3}$ in 1990 will be reduced to 47% by 2010. Unfortunately the same is not true according to the results of the photochemical simulations.

The effect from further reductions in $NO_x$ emissions is demonstrated by Figure 19. This figure shows the effect on average one-hour ozone concentrations from parametric scenarios of reducing the $NO_x$ emissions by 2010. These calculations show that, despite significant improvements of the European territory experiencing high one-hour ozone concentrations, there are still areas where the limit value will be exceeded.

## 7 Coupling Air Quality with Cost Data and Optimization of Scenarios

An important issue for policy makers is to establish the link between cost and environmental impact for each emission reduction scenario. A typical set of measures examined could be composed of scenarios for vehicle technology and inspection and maintenance (6 scenarios), fuel composition (16 scenarios) and non-technical measures (32 scenarios). Thus, the overall number of possible combinations covering the complete set of possible measures could rise easily to $6 \times 16 \times 32 = 3072$.

The cost data from most of these scenarios can be evaluated relatively easily. The same is not true for the environmental impact of these scenarios, especially for photochemical pollutants. A computer module has been developed at the later stages of AutoOil-1 (hence it could not be taken into consideration), which is based on grouping the emission effects into sets of scenarios with the same impact in each of the grid cells covering the domain of interest. Based on a predefined maximum tolerance limit, the large set of 3072 scenarios is grouped into classes with similar ranges of emission reduction effects. By the term maximum tolerance we in practice assume a range over which the following linear relationship holds for each of the grid cells of our domain. Hence, it is sufficient to carry out air quality modelling simulations for a limited number of scenarios which are characterized as class representatives in order to be able to characterize all possible combinations of scenarios for ozone:

$$\Delta[O_3] = f_1\Delta[CO] + f_2\Delta[NO_x] + f_3\Delta[VOC] \quad (1)$$

Naturally, the multiplication factors $f_1, f_2$ and $f_3$ are not known and need to be evaluated locally for each grid cell of the domain so that eqn. (1) can be used in evaluating $\Delta[O_3]$ for all scenarios where air quality modelling has not been carried out. By looking at the database of all available air quality simulations we had to examine all the emission combinations to locate three existing scenarios for which we had ozone simulations and which corresponded closest to $\Delta[CO]$, $\Delta[NO_x]$ and $\Delta[VOC]$ so that the predefined tolerance limit was satisfied. Then by solving these three equations the new expected ozone change can be calculated for each individual grid cell, even for scenarios for which direct air quality simulations are not available.

By this process, 'optimization for linearity' is achieved and it is possible to

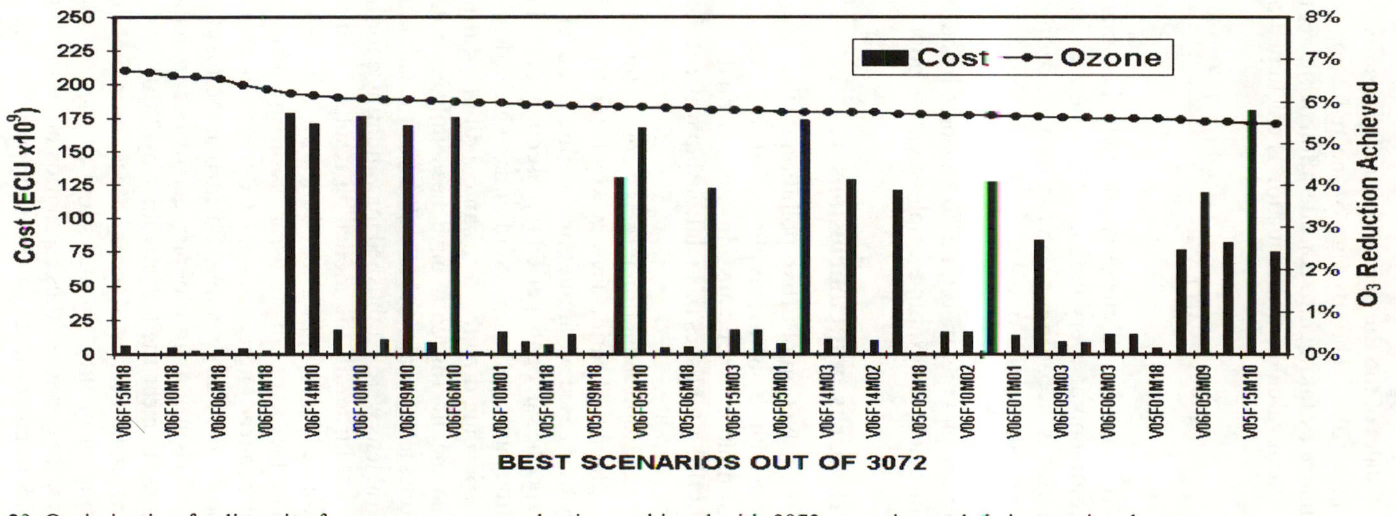


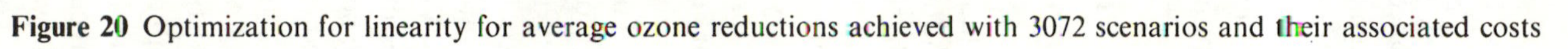

**Figure 20** Optimization for linearity for average ozone reductions achieved with 3072 scenarios and their associated costs

evaluate a large number of scenarios for cost and pollutant concentrations over the 15 states of the European Union. The results from this analysis are presented in Figure 20. The average ozone reduction over the 15 member states is calculated by taking into consideration ozone reductions in each individual cell. The 'mean value' of this reduction is calculated according to:

$$\overline{\Delta[O_3]} = \frac{1}{\text{cells}} \sum_{1}^{\text{cells}} \frac{[O_3]^{\text{scenario}} - [O_3]^{2010}}{[O_3]^{2010}} \tag{2}$$

For this reason, all the way through this analysis the cost and the ozone concentration at every individual cell were considered locally. The ranking of scenarios is according to the maximum ozone reduction achieved and the corresponding costs for achieving this are indicated by the bar chart. Clusters of scenarios with corresponding reductions are easily identified.

Similar plots as in Figure 20 can be carried out from other parameters of optimization, for example employment or population affected.

## 8 Remarks and Conclusions

For the reduction of urban pollution so that attainment within internationally imposed limits is accomplished, it is necessary to establish an accurate relationship between emissions and air quality. The methodology adopted for AutoOil-1 establishes this relationship while at the same time it tries to maintain physical reality.

The emission reduction targets were not linked to the strength of the emission through multiplication factors which treated atmospheric processes as static. They were based on simulations conducted under real meteorological conditions at a specific surrogate period representing annual mean values. The same is true for the anthropogenic conditions which dealt specifically with emissions corresponding directly to urban conditions during the surrogate period.

The monitoring data have been utilized both for the characterization of episodes for several cities as well as for the validation of air quality calculations. Several tests were established for validating enormous amounts of data matrices generated from all the modelling results. Furthermore, a procedure has been established for coupling cost data with achieved air quality reductions.

With the forecasting tools, emissions are predicted to decline significantly as a result of already agreed actions. Concentrations in all seven cities are estimated to be below both target values for carbon monoxide. For nitrogen dioxide, further reductions in emissions of nitrogen oxides are required of between 5% and 55% in order to meet the target values by 2010.

For ozone a series of scenarios was examined over a regional scale. Based on existing air quality objectives and the new World Health Organization guide values, the impact of these scenarios on future concentrations was analysed for the 15 member states of the European Union. National emissions of ozone precursors are predicted to be reduced by 35–40% over the period 1990 to 2010. However, this is not sufficient to meet any of the target values.

To meet the targets for one-hour average concentrations for ozone, the emissions need to be reduced by a further 40–60% from predicted 2010 levels. To meet the targets for the eight-hour average concentrations will require even larger emission reductions. In all the scenarios examined, including drastic emissions reductions, there was little predicted improvement in the eight-hour average concentrations and for the accumulated exposure over the 40 ppb value.

Based on the demonstrated results it could be concluded that, for meeting future air quality targets for ozone, an integrated strategy tackling all sources of precursor emissions needs to be taken into consideration, including those from stationary sources. The present status of the modelling tools and the recent computational advancements allow the extension of this methodology towards this direction.

## 9 Acknowledgements

The author would like to thank Dr C. Holman, Dr G. Kallos and Dr I. Ziomas for valuable discussions and contributions during the preparation of this article.

# Receptor Modeling for Air Quality Management

PHILIP K. HOPKE

## 1 Introduction

Since the early 1970s, when the need to improve air quality substantially was recognized and more stringent and effective regulations were put into place, there has been significant improvement in air quality with a concomitant benefit to public health. The US Environmental Protection Agency[1] has recently concluded a study of the costs and benefits to the US of implementing the Clean Air Act (CAA) and its amendments over the period of 1970 to 1990 and stated that the total cost of compliance with the CAA could be estimated to be 523 billion dollars, although major uncertainties in this value exist that could not be quantified. These expenditures resulted in a 40% decrease in the emissions of $SO_2$, a 30% decrease in the emissions of $NO_x$, a 45% reduction in volatile organic carbon (VOC) concentrations, a 50% reduction in CO, a 99% reduction in the emissions of lead and a 68% reduction in total suspended particulate matter (TSP). These reduced emissions, and thus decreased airborne concentrations, have resulted in a variety of benefits with the primary economic value in terms of improved human health, including longer productive life and fewer illnesses. These pollutant concentration changes also result in greater visibility, lower materials damage and ecosystem functioning. Calculating the benefits involves estimating both the effects and their associated costs, both of which have substantial uncertainties associated with them. However, the EPA reports that, with a 90% probability, the central estimate of the benefits lies between 10.5 and 40.6 trillion dollars. The mean of the distribution of benefits is 23 trillion dollars or an approximate benefit to cost ratio of 45.

However, recent epidemiological evidence has suggested that there are still substantial health effects that are being caused by levels of pollutants below the current levels of our National Ambient Air Quality Standards, and thus new regulations have been proposed for particulate matter and ozone.[2] The new regulations, if promulgated, will require more control of emissions and more

[1] Environmental Protection Agency, *The Benefits and Costs of the Clean Air Act, 1970 to 1990*, Draft Report to Congress, October 1996, Environmental Protection Agency, Washington.

[2] Environmental Protection Agency, announced on November 27, 1996, available from http://134.67.104.12/html/caaa/rules.htm.

concern for reductions of emissions in areas outside of the jurisdiction of the local air quality management authorities. Particulate matter will now also be regulated on the basis of small particle mass, particles in sizes less than 2.5 $\mu$m. Most of the mass of such particles is in the form of secondary particulate matter, particles formed by the conversion of gaseous emissions into lower vapor pressure substances that either homogeneously nucleate or condense onto the surface of existing particles. The concentrations of $O_3$ will have to be decreased by one third, but averaged over a longer time period (8 hours instead of 1 hour intervals). Such pollutants are transported over long distances and across political boundaries so that emission sources can affect locations quite far away. Thus, new approaches will be required to identify the sources of the pollutants and assess the effects of those sources on the ambient air quality. Receptor models have been effectively used in the past to identify local sources of particulate matter. In this chapter, the recent developments in the field and their possible application to these new challenges in air quality management will be presented.

## 2 Principle of Mass Balance

The fundamental principle of receptor modelling is that mass conservation can be assumed and a mass balance analysis can be used to identify and apportion sources of contaminants in the atmosphere. This methodology has generally been referred to within the air pollution research community as *receptor modelling*.[3,4] The approach to obtaining a data set for receptor modelling is to determine a large number of chemical constituents such as elemental concentrations in a number of samples. Alternatively, automated electron microscopy can be used to characterize the composition and shape of particles in a series of particle samples. In either case, a mass balance equation can be written to account for all $m$ chemical species in the $n$ samples as contributions from $p$ independent sources:

$$x_{ij} = \sum_{k=1}^{p} c_{ik} \cdot s_{kj} \tag{1}$$

where $x_{ij}$ is the $i$th elemental concentration measured in the $j$th sample, $c_{ik}$ is the gravimetric concentration of the $i$th element in material from the $k$th source, and $s_{kj}$ is the airborne mass concentration of material from the $k$th source contributing to the $j$th sample.

There exists a set of natural physical constraints on the system that must be considered in developing any model for identifying and apportioning the sources of airborne particle mass.[5] The fundamental, natural physical constraints that must be obeyed are:

1. The original data must be reproduced by the model; the model must explain the observations.
2. The predicted source compositions must be non-negative; a source cannot have a negative percentage of an element.

[3] P. K. Hopke, *Receptor Modeling in Environmental Chemistry*, Wiley, New York, 1985.

[4] *Receptor Modeling for Air Quality Management*, ed. P. K. Hopke, Elsevier, Amsterdam, 1991.

[5] R.C. Henry, in *Receptor Modeling for Air Quality Management*, ed. P. K. Hopke, Elsevier, Amsterdam, 1991, pp. 117–147.

3. The predicted source contributions to the aerosol must all be non-negative; a source cannot emit negative mass.
4. The sum of the predicted elemental mass contributions for each source must be less than or equal to total measured mass for each element; the whole is greater than or equal to the sum of its parts.

While developing and applying these models, it is necessary to keep these constraints in mind in order to be certain of obtaining physically realistic solutions.

## 3 Currently Used Receptor Models

Most of the previous applications of receptor modeling have been to airborne particulate matter. The US EPA has approved the use of one model, the chemical mass balance (CMB) model, as part of the implementation planning process.[6] This model assumes that the number of sources and their compositional profiles are known and that the only remaining unknown in eqn. (1) is the mass contribution of each source to each sample. Such an approach works well for larger sized particles ($> 1\ \mu$m), where there are generally sufficient compositional differences between emissions from different sources to permit them to be distinguished from one another. The CMB model is a multiple regression model. In the implementation prepared for the US EPA, the model incorporates the estimated uncertainties in the emitted source material compositions as well as the uncertainties in the measured elemental compositions of the particulate matter samples.[7] The effective variance least squares fitting procedure is used to solve the problem iteratively.[8,9] Recent applications of the CMB method to airborne particulate matter are summarized in Table 1.

Source apportionment studies have typically been carried out and reported from $PM_{10}$ non-attainment areas, most of which are in the western US. Thus, only a few areas in the eastern US (*e.g.* the Stuebenville area of Ohio, and the Granite City and Chicago areas of Illinois) report source contribution estimates. Since there are significant climatological differences between the arid west and the more moist east, the utility of the CMB methodology to resolve sources across the entire US has yet to be fully tested.

Some of the studies did not measure major chemical species that constitute suspended particle mass, *i.e.* elements, sulfate, nitrate, ammonium and carbon. For example, studies[10,11] of particle mass in south Chicago did not measure

[6] Environmental Protection Agency, *$PM_{10}$ SIP Development Guideline*, EPA Report EPA-450/2-86-001, US Environmental Protection Agency, Office of Air Quality Planning and Standards, Research Triangle Park, NC, 1987.

[7] J. G. Watson, N. F. Robinson, J. C. Chow, R. C. Henry, B. M. Kim, T. G. Pace, E. L. Meyer and Q. Nguyen, *Environ. Software*, 1990, **5**, 38.

[8] J. A. Cooper, J. G. Watson and J. J. Huntzicker, *Atmos. Environ.*, 1984, **18**, 1347.

[9] M. D. Cheng, P. K. Hopke and D. Jennings, *Chemom. Intell. Lab. Syst.*, 1988, **4**, 239.

[10] P. K. Hopke, W. Wlaschin, S. Landsberger, C. Sweet and S. J. Vermette, in *PM-10: Implementation of Standards*, ed. C. V. Mathai and D. H. Stonefield, Air Pollution Control Association, Pittsburgh, 1988, pp. 484–494.

[11] S. J. Vermette, A. L. Williams and S. Landsberger, in *$PM_{10}$ Standards and Nontraditional Particulate Source Controls*, ed. J. C. Chow and D. M. Ono, Air & Waste Management Association, Pittsburgh, 1992, vol. 1, pp. 262–278.

**Table 1** Receptor model source contributions to $PM_{10}$*

| *Sampling site* | *Ref.* | *Time period* | *Primary geological* | *Primary construction* | *Primary motor vehicle exhaust* | *Primary vegetative burning* | *Secondary ammonium sulfate* | *Secondary ammonium nitrate* | *Misc. source 1* | *Misc. source 2* | *Misc. source 3* | *Misc. source 4* | *Measured $PM_{10}$ concentration* |
|---|---|---|---|---|---|---|---|---|---|---|---|---|---|
| Central Phoenix, AZ | 39 | Winter 1989–1990 | 33.0 | 0.0 | 25.0 | 2.3 | 0.2 | 2.8 | 0.0 | 0.0 | 0.0 | 0.0 | 64.0 |
| Corona de Tucson, AZ | 14 | Winter 1989–1990 | 17.0 | 0.0 | 1.6 | 0.0 | 1.9 | 0.0 | 0.0 | 0.0 | 0.0 | 19.1 | |
| Craycroft, AZ | 14 | Winter 1989–1990 | 13.0 | 0.0 | 8.3 | 0.0 | 0.7 | 0.6 | 1.2[a] | 0.0 | 0.0 | 0.0 | 23.4 |
| Downtown Tucson, AZ | 14 | Winter 1989–1990 | 26.0 | 5.1 | 14.0 | 0.0 | 1.0 | 0.2 | 1.3[a] | 0.0 | 0.0 | 0.0 | 48.0 |
| Hayden 1, AZ (Garfield) | 40 | 1986 | 5.0 | 2.0[b] | 0.0 | 0.0 | 4.0 | 0.0 | 74.0[c] | 5.0[d] | 1.0[e] | 0.0 | 105.0 |
| Hayden 2, AZ (Jail) | 40 | 1986 | 21.0 | 4.0[b] | 0.0 | 0.0 | 4.0 | 0.0 | 28.0[c] | 0.0 | 1.0[e] | 0.0 | 59.0 |
| Orange Grove, AZ | 14 | Winter 1989–1990 | 20.0 | 0.0 | 15.0 | 0.0 | 0.4 | 0.4 | 0.0 | 0.0 | 0.0 | 0.0 | 34.2 |
| Phoenix, AZ (Estrella Park) | 39 | Winter 1989–1990 | 37.0 | 0.0 | 10.0 | 0.9 | 1.6 | 0.0 | 0.0 | 0.0 | 0.0 | 0.0 | 55.0 |
| Phoenix, AZ (Gunnery Rg.) | 39 | Winter 1989–1990 | 20.0 | 0.0 | 5.5 | 0.0 | 1.0 | 0.0 | 0.0 | 0.0 | 0.0 | 0.0 | 27.0 |
| Phoenix, AZ (Pinnacle Park) | 39 | Winter 1989–1990 | 7.0 | 0.0 | 2.9 | 1.0 | 0.9 | 0.0 | 0.0 | 0.0 | 0.0 | 0.0 | 12.0 |
| Rillito, AZ | 41 | 1988 | 42.7 | 13.8[b] | 1.2[f] | 0.0 | 0.0 | 0.0 | 11.6[g] | 0.0 | 0.0 | 0.0 | 79.5 |
| Scottsdale, AZ | 39 | Winter 1989–1990 | 25.0 | 0.0 | 19.0 | 7.4 | 0.6 | 3.6 | 0.0 | 0.0 | 0.0 | 0.0 | 55.0 |
| West Phoenix, AZ | 39 | Winter 1989–1990 | 30.0 | 0.0 | 25.0 | 10.0 | 0.4 | 3.1 | 0.0 | 0.0 | 0.0 | 0.0 | 69.0 |
| Anaheim, CA | 42 | 1986 | 21.2 | 0.0 | 4.1[i] | 0.0 | 7.0 | 9.8 | 0.4[j] | 1.4[h] | 8.2[k] | 0.0 | 52.1 |
| Anaheim, CA (Summer) | 43 | Summer 1987 | 11.4 | 0.0 | 8.5 | 0.0 | 9.0 | 2.9 | 0.0[j] | 6.5[h] | 0.0 | 0.0 | 51.3 |
| Anaheim, CA (Fall) | 43 | Fall 1987 | 13.2 | 0.0 | 37.2 | 0.0 | 3.7 | 38.5 | 0.0[j] | 3.1[h] | 0.0 | 0.0 | 104.0 |
| Azusa, CA (Summer) | 43 | Summer 1987 | 34.9 | 0.0 | 15.9 | 0.0 | 11.4 | 6.1 | 0.0[j] | 5.7[h] | 0.0 | 0.0 | 92.1 |
| Bakersfield, CA | 44 | 1986 | 27.4 | 3.0 | 5.5 | 9.6[l] | 5.6 | 0.0 | 0.5[j] | 0.0 | 0.0 | 0.0 | 67.6 |
| Bakersfield, CA | 45 | 1988–1989 | 42.9 | 1.6 | 7.7 | 6.5 | 5.5 | 12.7 | 1.0[m] | 1.5[n] | 0.6[k] | 0.0 | 79.6 |
| Burbank, CA | 42 | 1986 | 21.3 | 0.0 | 6.1[i] | 0.0 | 7.2 | 10.2 | 0.1[j] | 0.9[h] | 9.8[k] | 0.0 | 56.6 |
| Burbank, CA (Summer) | 43 | Summer 1987 | 14.0 | 0.0 | 17.0 | 0.0 | 12.4 | 6.5 | 0.0[j] | 5.7[h] | 0.0 | 0.0 | 72.3 |
| Burbank, CA (Fall) | 43 | Fall 1987 | 11.0 | 0.0 | 39.1 | 0.0 | 3.1 | 25.1 | 0.0[j] | 1.9[h] | 0.0 | 0.0 | 94.8 |
| Chula Vista 1, CA (Bayside) | 46 | 1986 | 6.7 | 0.0 | 0.8 | 0.0 | 7.5 | 0.0 | 0.4[j] | 2.7[h] | 2.0[k] | 0.0 | 28.8 |
| Chula Vista 2, CA (Del Ray) | 46 | 1986 | 8.2 | 0.3 | 1.5 | 0.0 | 8.9 | 0.0 | 0.6[j] | 1.8[h] | 0.0 | 0.0 | 31.1 |
| Chula Vista 3, CA | 46 | 1986 | 9.7 | 0.3 | 1.4 | 0.0 | 8.2 | 0.0 | 0.6[j] | 1.7[h] | 0.0 | 0.0 | 29.6 |
| Claremont, CA (Summer) | 43 | Summer 1987 | 19.4 | 0.0 | 14.4 | 0.0 | 9.5 | 6.3 | 0.0[j] | 4.7[h] | 0.0 | 0.0 | 70.0 |
| Crows Landing, CA | 45 | 1988–1989 | 32.2 | 0.0 | 2.2 | 3.4 | 2.8 | 6.5 | 0.5[m] | 1.5[n] | 1.2[k] | 0.0 | 52.5 |
| Downtown Los Angeles, CA | 42 | 1986 | 23.8 | 0.0 | 6.4[i] | 0.0 | 7.6 | 11.2 | 0.0 | 1.3[h] | 7.9[k] | 0.0 | 60.2 |
| Downtown Los Angeles, CA (Summer) | 43 | Summer 1987 | 12.7 | 0.0 | 16.2 | 0.0 | 13.0 | 4.4 | 0.0[j] | 6.5[h] | 0.0 | 0.0 | 67.6 |
| Downtown Los Angeles, CA (Fall) | 43 | Fall 1987 | 9.4 | 0.0 | 41.1 | 0.0 | 3.9 | 27.5 | 0.0[j] | 1.8[h] | 0.0 | 0.0 | 98.6 |
| Fellows, CA | 45 | 1988–1989 | 29.0 | 1.4 | 2.1 | 3.4 | 5.1 | 7.5 | 7.0[m] | 1.4[n] | 1.4[k] | 0.0 | 54.6 |
| Fresno, CA | 44 | 1986 | 17.1 | 0.7 | 4.0 | 9.2[l] | 1.8 | 0.0 | 0.1[j] | 0.0 | 0.0 | 0.0 | 48.1 |
| Fresno, CA | 45 | 1988–1989 | 31.8 | 0.0 | 6.8 | 5.1 | 3.6 | 10.4 | 0.3[m] | 1.0[n] | 0.1[k] | 0.0 | 71.5 |

| | | | | | | | | | | | | | |
|---|---|---|---|---|---|---|---|---|---|---|---|---|---|
| Hawthorne, CA (Summer) | 43 | Summer 1987 | 7.5 | 0.0 | 5.6 | 0.0 | 15.0 | 0.6 | 0.0[j] | 7.0[h] | 0.0 | 0.0 | 45.9 |
| Hawthorne, CA (Fall) | 43 | Fall 1987 | 8.9 | 0.0 | 35.1 | 0.0 | 5.1 | 20.4 | 0.0[j] | 3.7[h] | 0.0 | 0.0 | 85.1 |
| Indio, CA | 47 | | 33.0 | 3.0 | 4.4 | 7.1 | 3.6 | 4.1 | 0.2[j] | 1.0[h] | 0.0 | 0.0 | 58.0 |
| Kern Wildlife Refuge, CA | 45 | 1988–1989 | 15.1 | 2.0 | 2.2 | 4.0 | 3.3 | 1.5 | 0.5[m] | 1.5[n] | 0.7[k] | 0.0 | 47.8 |
| Lennox, CA | 42 | 1986 | 16.0 | 0.1 | 4.6[i] | 0.0 | 7.6 | 7.9 | 0.2[j] | 3.1[h] | 7.6[k] | 0.0 | 46.9 |
| Long Beach, CA | 43 | 1986 | 20.7 | 0.0 | 5.1[i] | 0.0 | 8.0 | 9.2 | 0.1[j] | 2.0[h] | 6.4[k] | 0.0 | 51.9 |
| Long Beach, CA (Summer) | 43 | Summer 1987 | 11.1 | 0.0 | 6.3 | 0.0 | 10.9 | 0.8 | 0.1[j] | 2.2[h] | 0.0 | 0.0 | 46.1 |
| Long Beach, CA (Fall) | 43 | Fall 1987 | 11.3 | 0.0 | 42.8 | 0.0 | 3.8 | 23.2 | 0.0[j] | 2.7[h] | 0.0 | 0.0 | 96.1 |
| Magnolia, CA | 48 | 1988 | 31.7 | 0.0 | 11.2 | 0.0 | 4.9 | 19.7 | 0.3[j] | 1.2[h] | 1.2[o] | 0.0 | 66.0 |
| Palm Springs, CA | 47 | | 16.4 | 1.4 | 2.3 | 5.1 | 3.7 | 4.2 | 0.1[j] | 0.5[h] | 0.0 | 0.0 | 35.1 |
| Riverside, CA | 48 | 1988 | 32.6 | 0.0 | 7.0 | 0.0 | 4.8 | 21.4 | 0.3[j] | 1.3[h] | 1.1[o] | 0.0 | 64.0 |
| Rubidoux, CA | 42 | 1986 | 43.1 | 4.0[j] | 5.6[i] | 0.0 | 6.4 | 21.3 | 0.3[j] | 1.0[h] | 5.9[k] | 0.0 | 87.4 |
| Rubidoux, CA (Summer) | 43 | Summer 1987 | 34.9 | 4.5 | 17.3 | 0.0 | 9.5 | 27.4 | 0.0[j] | 5.1[h] | 0.0 | 0.0 | 114.8 |
| Rubidoux, CA (Fall) | 43 | Fall 1987 | 19.2 | 16.1 | 30.3 | 0.0 | 2.1 | 31.6 | 0.0[j] | 1.1[h] | 0.0 | 0.0 | 112.0 |
| Rubidoux, CA | 48 | 1988 | 48.0 | 0.0 | 10.2 | 0.0 | 5.3 | 21.7 | 0.4[j] | 1.5[h] | 5.7[o] | 0.0 | 87.0 |
| San Nicolas Island, CA (Summer) | 43 | Summer 1987 | 1.6 | 0.0 | 0.9 | 0.0 | 3.7 | 0.5 | 0.0[j] | 4.3[h] | 0.0 | 0.0 | 17.4 |
| Stockton, CA | 45 | 1989 | 34.4 | 0.5 | 5.2 | 4.8 | 3.1 | 7.0 | 0.7[m] | 1.8[n] | 0.0[k] | 0.0 | 62.4 |
| Upland, CA | 42 | 1986 | 25.4 | 0.4[j] | 4.1[i] | 0.0 | 6.4 | 14.5 | 0.6[j] | 0.6[h] | 7.8[k] > | 0.0 | 58.0 |
| Telluride 1, CO (Central) | 49 | Winter 1986 | 32.0 | 0.0 | 0.0 | 98.7 | 0.0 | | 61.3[p] | 0.0 | 0.0 | 0.0 | 208.0 |
| Telluride 2, CO (Society Turn) | 49 | Winter 1986 | 12.1 | 0.0 | 0.0 | 7.3 | 0.0 | | 7.3[p] | 0.0 | 0.0 | 0.0 | 27.0 |
| Pocatello, ID | 50 | 1990 | 8.3 | 7.5[q] | 0.1 | 0.0 | 0.0 | 0.0 | 0.0 | 0.0 | 84.1[r] | 0.0 | 100.0 |
| S. Chicago, IL | 10 | 1986 | 27.2 | 2.4 | 2.8 | 0.0 | 15.4[s] | | 15.1[t] | 2.2[u] | 0.0 | 0.0 | 80.1 |
| S.E. Chicago, IL | 11 | 1988 | 14.7[v] | 0.0 | 0.9[f] | 0.0 | 7.7 | | 0.8[t] | 0.3[h] | 1.1[w] | 7.7[g] | 41.0 |
| Reno, NV (Non-sweeping) | 51 | Winter 1987 | 9.7 | 0.0 | 8.7 | 0.1 | 0.6 | 0.2 | 0.0 | 0.0 | 0.0 | 0.0 | 20.4 |
| Reno, NV (Sweeping) | 51 | Winter 1987 | 11.8 | 0.0 | 11.0 | 1.2 | 0.8 | 0.2 | 0.0 | 0.0 | 0.0 | 0.0 | 24.9 |
| Reno, NV | 13 | 1986–1987 | 14.9 | 0.0 | 10.0 | 1.9 | 1.3 | 0.6 | 0.0 | 0.0 | 0.0 | 0.0 | 30.0 |
| Sparks, NV | 13 | 1986–1987 | 15.1 | 0.0 | 11.6 | 13.4 | 2.7 | 0.9 | 0.0 | 0.0 | 0.2[k] | 0.0 | 41.0 |
| Verdi, NV | 13 | 1986–1987 | 7.8 | 0.0 | 4.0 | 1.1 | 0.9 | 0.1 | 0.0 | 0.0 | 0.0 | 0.0 | 15.0 |
| Follansbee, OH | 52 | 1991 | 10.0 | 0.0 | 35.0 | 0.0 | 16.0 | | 9.3[t] | 0.0 | 0.0 | 0.0 | 66.0 |
| Mingo, OH | 52 | 1991 | 12.0 | 0.0 | 14.0 | 4.1 | 15.0 | | 3.4[t] | 11.0[x] | 0.0 | 0.0 | 60.0 |
| Sewage Plant, OH | 52 | 1991 | 22.0 | 0.0 | 12.0 | 0.0 | 13.0 | | 6.6[t] | 8.7[x] | 0.0 | 0.0 | 62.0 |
| Steubenville, OH | 52 | 1991 | 8.3 | 0.0 | 14.0 | 0.8 | 14.0 | | 3.8[t] | 5.0[x] | 0.0 | 0.0 | 46.0 |
| WTOV Tower, OH | 52 | 1991 | 7.4 | 0.0 | 16.0 | 0.2 | 15.0 | | 3.4[t] | 7.9[x] | 0.0 | 0.0 | 49.0 |

*In $\mu g\,m^{-3}$.
[a]Smelter background aerosol. [b]Cement plant sources, including kiln stacks, gypsum pile, and kiln area. [c]Copper ore. [d]Copper tailings. [e]Copper smelter building. [f]Heavy-duty diesel exhaust emission. [g]Background aerosol. [h]Marine aerosol, road, salt, and sea salt plus sodium nitrate. [i]Motor vehicle exhaust from diesel and leaded gasoline. [j]Residual oil combustion. [k]Secondary organic carbon. [l]Biomass burning. [m]Primary crude oil. [n]NaCl + $NaNO_3$. [o]Lime. [p]Road sanding material. [q]Asphalt industry. [r]Phosphorus/phosphate industry. [s]Regional sulfate. [t]Steel mills. [u]Refuse incinerator. [v]Local road dust, coal yard road dust, steel haul road dust. [w]Incineration. [x]Unexplained mass.

carbon. For this reason, the sum of source contribution estimates is much lower than the measured mass. In many cases, samples and associated analyses are performed for other purposes and then subsequently used for receptor modelling even if they do not provide the full set of information needed to complete the analysis. Samples selected for chemical analysis are often biased toward the highest $PM_{10}$ mass concentrations in these studies, so average source contribution estimates are probably not representative of annual averages.

Many of these studies were conducted during the late 1980s, when a portion of the vehicle fleet still used leaded gasoline. While the lead and bromine in motor vehicle emissions facilitated the distinction of motor vehicle contributions from other sources, they were also associated with higher emission rates than vehicles using unleaded fuels. Lead has been removed from currently used vehicle fuels.

Uncertainty estimates of source contribution estimates are not usually reported with the average values summarized in Table 1. Standard errors are calculated in source apportionment studies, and typically range from 15 to 30% of the source contribution estimate. They are much higher when the chemical source profiles for different sources are highly uncertain or too similar to distinguish one source from another. A major problem is the difficulty in assessing the errors when source sampling in the affected area cannot be done as part of the study. Glover *et al.*[12] find a significant improvement in the quality of the analyses when source profiles can be determined for the sources identified and analysed as part of the study. It is quite difficult to estimate the errors that can be produced by source profile misspecification that may occur when source profile libraries are used rather than directly measured emission compositions. Multivariate calibration methods described in a subsequent section can assist in determining the suitability of library source profiles, but it represents a significant weakness of the CMB approach.

Different measurement sites within the same airshed show different proportions of contributions from the same sources. Most often, the sites in close proximity to an emitter show a much larger contribution from that emitter than sites that are distant from that emitter, even by distances as low as 10 km (*e.g.* Chow *et al.*[13,14]). Given the mass, elements, ion and carbon components measured on source and receptor samples in most of the studies in Table 1, greater differentiation among sources (*e.g.* diesel and gasoline vehicle exhaust, meat cooking and other organic carbon sources, different sources of fugitive dust, and secondary aerosol precursors) is not possible. Despite these limitations, source apportionment modelling has been useful in identifying the major sources that need to be reduced in order to yield $PM_{10}$ attainment.

Because of the separation of emitted primary particles from the secondary particle precursor gaseous emissions, materials like $SO_4^{2-}$, $NO_3^-$, $NH_4^+$ and

[12] D. M. Glover, P. K. Hopke, S. J. Vermette, S. Landsberger and D. R. D'Auben, *J. Air Waste Manage. Assoc.*, 1991, **41**, 294.

[13] J. C. Chow, J. G. Watson, C. A. Frazier, R. T. Egami, A. Goodrich and C. Ralph, in *PM-10: Implementation of Standards*, ed. C. V. Mathai and D. H. Stonefield, Air Pollution Control Association, Pittsburgh, 1988, pp. 438–457.

[14] J. C. Chow, J. G. Watson, D. H. Lowenthal, C. A. Frazier, B. A. Hinsvark, L. C. Prichett and G. R. Neuroth, in *$PM_{10}$ Standards and Nontraditional Particular Source Controls*, ed. J. C. Chow and D. M. Ono, Air & Waste Management Association, Pittsburgh, 1992, vol. 1, pp. 231–243.

secondary organic particles are not directly attributed to source categories. These materials will need to be controlled in order to meet a fine particle standard, and thus it is essential to be able to associate the observed concentrations of secondary species with specific source types and potentially with specific source locations. Only then can an effective implementation plan be developed. Since the primary particle source signatures are lost in transit, other kinds of information need to be included in order to identify and apportion sources of secondary particulate matter.

## 4 Secondary Aerosol Mass

As mentioned previously, none of the methods described before can provide a definite indication of the sources of secondary particles such as sulfate, nitrate or secondary organic materials. The usual results of CMB analysis is to list 'sulfate' as a source or possibly describe it as 'regional sulfate'. Similar results are typically obtained through factor analysis. In order really to develop effective control strategies, it will be necessary to attribute the secondary particle mass to the original gaseous precursor sources. In order to make such an apportionment, additional information must be included in the analysis. This information is generally in the form of spatial/temporal information or in terms of meteorology as defined by air parcel back trajectories.

### *Spatial Analysis*

The basic mathematical framework of this method is the same as a principal components analysis. However, instead of examining the variation of a number of measured species in samples at a single site, the input data here are the values of a single variable measured at a variety of sites at multiple times. Thus, the analysis is seeking spatial and temporal variations of the measured variable within the data set rather than the interrelationships among the variables. This approach has been used to interpret the $SO_2$ concentrations over St. Louis, Missouri,[15] and for particulate sulfur concentrations in the western US.[16] Malm *et al.*[17] used EOF in addition to a trajectory based method called Area of Influence Analysis. These results will be presented in the next section. Henry *et al.*[18] used a modified EOF analysis to look for $SO_2$ sources over the southwestern US based on 3-day-long samples from the National Park Service sampling network. The results of this analysis are shown in Figure 1. In this analysis, areas of high positive values are likely source areas while negative values represent regions that serve as sulfur sinks. It is clear that several well-known sources are identified. From the results in this figure, it is also possible to make some simple preliminary quantitative estimates of sulfur dioxide oxidation rates and sulfate deposition velocities. Thus, this approach is promising in areas where such widespread sampling and analyses programs that could address regional scale production

[15] J.T. Peterson, *Atmos. Environ.*, 1970, **4**, 501.

[16] L.L. Ashbaugh, L.O. Myrup and R.G. Flocchini, *Atmos. Environ.*, 1984, **18**, 783.

[17] W.C. Malm, K.A. Gebhart and R.C. Henry, *Atmos. Environ.*, 1990, **24A**, 3047.

[18] R.C. Henry, Y.J. Wang and K.A. Gebhart, *Atmos. Environ.*, 1991, **25A**, 503.

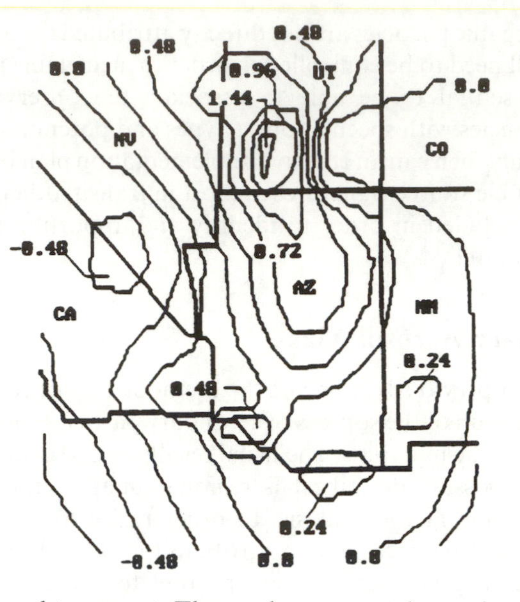


**Figure 1** Average balance of sources minus sinks of airborne sulfate-sulfur for the period July 1983 and June 1985. Units are $10^{-3}\,\mathrm{ng\,m^{-3}\,s^{-1}}$. (Taken from Henry *et al.*[18] and used with permission)

and transport. Thus, other approaches to incorporating additional information beside the sample's chemical composition need to be employed.

## *Methods Incorporating Back Trajectories*

Dispersion models describe the transport of the particles from a source to the sampling location. However, using an analogous model of atmospheric transport, a trajectory model calculates the position of the air being sampled backward in time from the receptor site from various starting times throughout the sampling interval. The trajectories are then used in Residence Time Analysis, Areas of Influence Analysis, Quantitative Bias Trajectory Analysis, Potential Source Contribution Function and Residence Time Weighted Concentrations. Recently, PSCF has been extended to yield an apportionment by combining it with the emission inventory for the area being modeled. Thus, these methods now appear to be such that they can be applied to identifying locations of emissions and possibly to the apportioning of secondary aerosol mass to those identified locations.

*Residence Time Analysis.* An initial effort was made by Ashbaugh[19] to make use of air parcel back trajectories to identify likely source locations for particulate sulfur observed at the Grand Canyon. A gridded array is created around the sampling location. Trajectories are a sequence of segments, each of which represents a fixed amount of time. Thus, each endpoint can be considered to be an indication that the air parcel has spent a given time within that grid cell. The total 'residence time' that air spends in the given cells would be the total number of endpoints that fall into that cell. These values can be plotted over a map. The residence time values associated with high or low concentration can be plotted to

[19] L. L. Ashbaugh, *J. Air. Pollut. Contr. Assoc.*, 1983, **33**, 1096.

examine likely directions from which contaminated or clean air is transported to the sampling site.

The problem with this method is that all of the trajectories begin at the receptor site and thus the residence time is maximum in cells surrounding the sampling location. Ashbaugh *et al.*[20] suggest one solution to this problem that will be described below. An alternative method, which has come to be called Residence Time Analysis, was developed by Poirot and Wishinski.[21] In their method, they first interpolate along each trajectory segment to estimate the fraction of time spent in each grid cell and then sum the residence time for that cell. They propose a method to adjust the resulting grid cell values for the geometrical problem of high values in the region immediately adjacent to the receptor site.

Figure 2 illustrates their approach to calculating a correction factor. They consider an annular ring of width equal to that of a grid cell at a distance $D_{ij}$ from the receptor site. The area of this annulus can then be compared to that of a grid cell.

$$\frac{\text{Annular area}}{\text{Single grid cell area}} = \frac{\pi[D_{ij} + L)^2 - (D_{ij}\ L)^2]}{L^2} \tag{2}$$

where $L$ is the half-width of a grid cell. This ratio increases as the distance increases so that its inverse will represent the decrease in the probability that a grid cell will be traversed by a trajectory compared with a centrally located cell. Thus, we can consider the right-hand term of eqn. (2) can be defined to be the geometric adjustment factor $F_{ij}$. The residence time for any specific grid cell is then multiplied by $F_{ij}$ to yield the geometrically corrected residence time.

Poirot and Wishinski[21] applied this method to the source locations for sulfur arriving at Underhill, Vermont, during 1978–1980. The resulting residence time analysis map for high sulfate summer days is shown in Figure 3. The influence of the Ohio River Valley and the region around the Great Lakes show the highest values, in good agreement with the known presence of $SO_2$ sources in these areas.

*Areas of Influence Analysis.* Areas of Influence Analysis (AIA) is an extension of residence time analysis in which back trajectories are used to estimate the amount of time an air parcel spends in the areas of sources. The method was applied to particular sulfate in the southwestern US by Malm *et al.*[17] in conjunction with an EOF analysis. The method begins by identifying those 'extreme' samples that have values which are at least 1 standard deviation above the median value at each measurement site. The region was divided into 1° latitude by 1° longitude. For each cell, the extreme concentration residence time is defined as the number of trajectory endpoints associated with extreme valued samples. Since all of the trajectories emanated from the receptor site, the extreme valued residence times will be highest around that site. In order to eliminate this central tendency, a new function can be defined by dividing the residence time value by an equal probability residence time surface. This equal probability surface was developed by assuming that an air parcel can arrive at the receptor from any direction with equal probability and that the wind speed associated with each parcel is constant.

[20] L. L. Ashbaugh, W. C. Malm and W. D. Sadeh, *Atmos. Environ.*, 1985, **19**, 1263.

[21] R. L. Poirot and P. R. Wishinski, *Atmos. Environ.*, 1986, **20**, 1457.

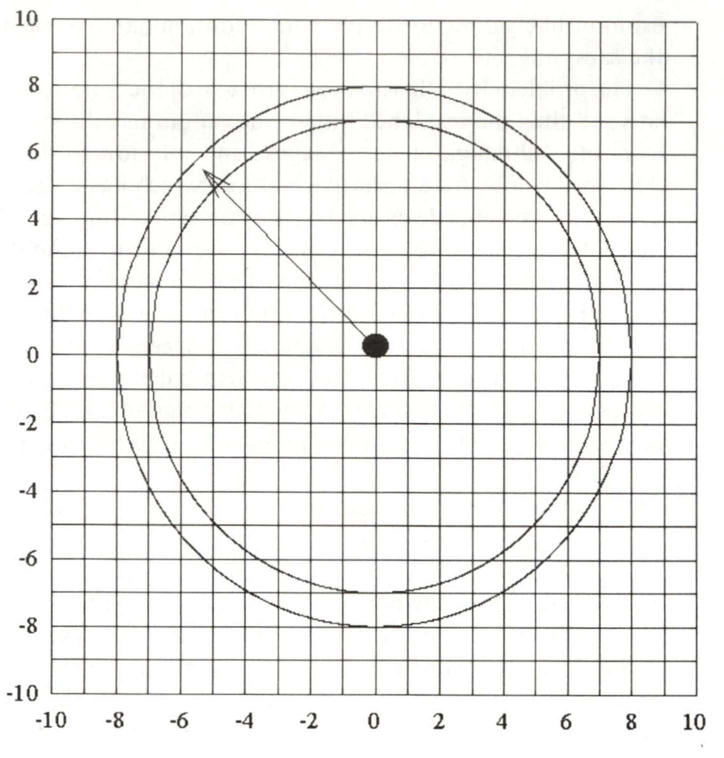


**Figure 2** Illustration of the geometrical correction factor for correcting the residence time for a given grid cell a distance $D_{ij}$ from the receptor site

This new function, the extreme source contribution function (ESCF), would yield an estimate of the relative contributions from the different source cells to the extreme concentrations observed at each receptor site.

In AIA, cells in the known source regions are identified. Then, for each receptor site, an average value of the ESCF can be calculated from the collection of cell values from all of the receptor sites. These cell values can then be plotted on a single map for the region to show where the collected particulate matter is likely to have originated. The EOF results can be plotted on the same map so that they can be compared and complement each other in providing insights into the source/receptor relationships.

Figure 4 shows the results for the area of southern California. These results show that this region has a strong influence on Joshua Tree and Death Valley National Monuments in southern California, the Grand Canyon and Bryce Canyon National Parks in Utah–Arizona and the Great Basin National Park in Nevada.

Similar analyses have been performed for a series of locations including Mexico particularly around Monterrey, power plant areas in northern Arizona near the border with Utah (four corners regions), copper smelter areas in southern Arizona, the Texas Gulf Coast region, the El Paso, Texas area, several locations in the northwestern US (Washington–Oregon and San Francisco), the northern midwestern US and the southern Idaho–northern Utah region. This

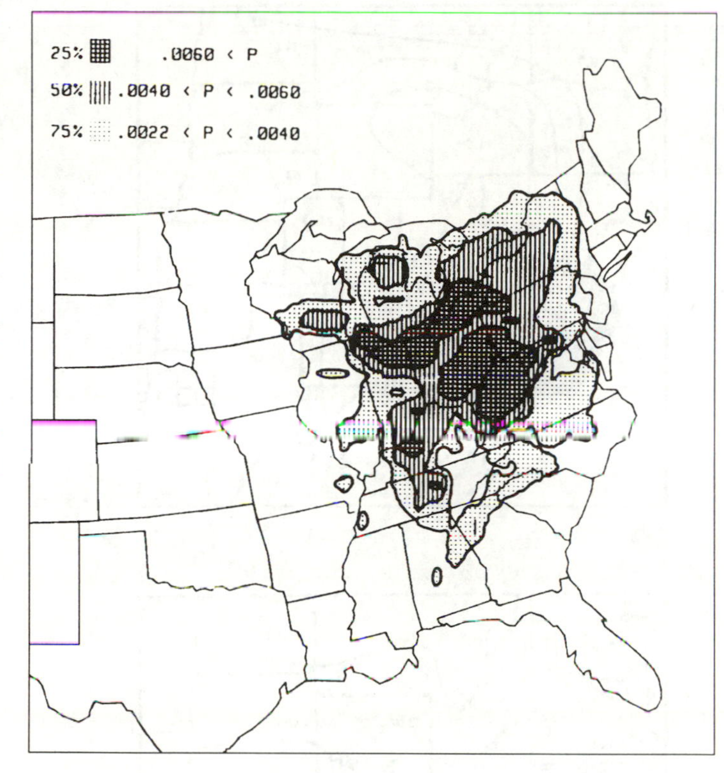


**Figure 3** Residence time probabilities for all high-sulfate summer days during 1978, 1979 and 1980 ($[SO_4^{2-}] > 10\,\mu g\,m^{-3}$]. 25% of the residence time hours are contained within each separately shaded area. $P$ equals the probability of residence in an 80 km × 80 km grid square within the shaded area. (Taken from Poirot and Wishinski[21] and used with permission)

technique again appears to work well in this area because of the nature of the sampling network and the other factors discussed by Henry *et al.*[18] and described above.

*Quantitative Bias Trajectory Analysis.* A more sophisticated approach is Quantitative Transport Bias Analysis.[22,23] This approach explicitly incorporates deposition along the back trajectory. The transport potential for a given sampling time includes the mean transport computed using the trajectory model and the uncertainty imposed by the atmospheric dispersion that occurs on route. The probability of a reactive depositing tracer arriving at a point $x$ at time $t$ is given by Lamb and Seinfeld[24] and Cass[25] as

$$A(x,t) = \int_{t-\tau}^{t} \int_{-\infty}^{\infty} \int_{-\infty}^{\infty} T(x,t) \mid x',t')\mathrm{d}x'\mathrm{d}t' \tag{3}$$

where $T(x,t \mid x',t')$ is the potential mass transfer function. It is given by

$$T(x,t \mid x',t') = Q(x,t \mid x',t')R(t \mid t')D(x',t')\Lambda(x',t') \tag{4}$$

[22] G. J. Keeler, Ph.D. Dissertation, University of Michigan, Ann Arbor, 1987.
[23] G. J. Keeler and P. J. Samson, *Environ. Sci. Technol.*, 1989, **23**, 1358.
[24] R. G. Lamb and J. H. Seinfeld, *Environ. Sci. Technol.*, 1973, **7**, 253.
[25] G. R. Cass, *Atmos. Environ.*, 1981, **15**, 1227.

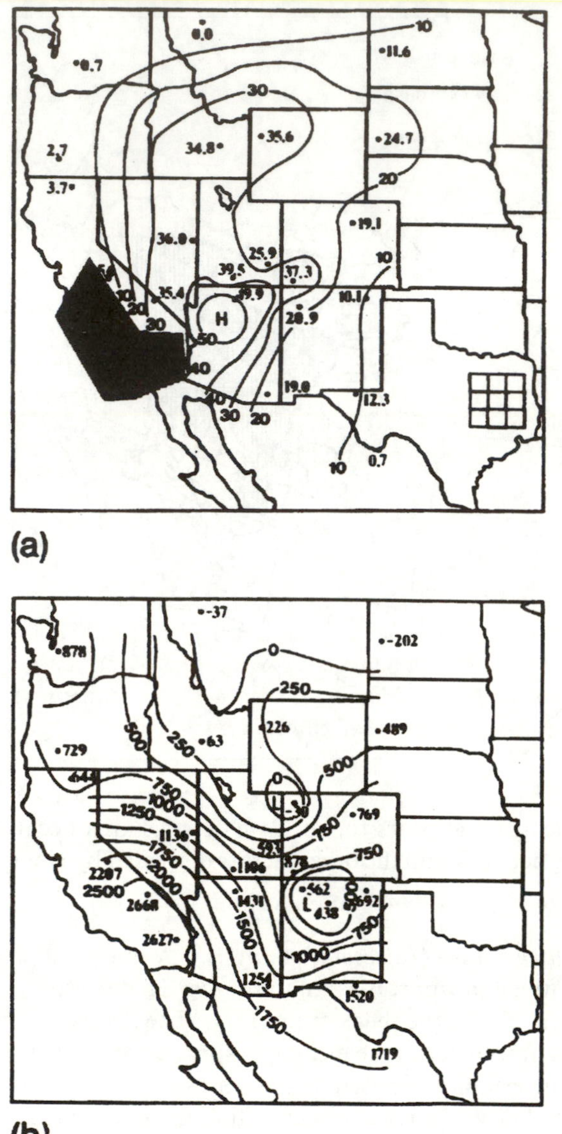


**Figure 4** Area of influence analysis results for southern California on the areas of the southwestern US shown at the top. The rotated principal components analysis results are shown below. (Taken from Malm *et al.*[17] and used with permission)

in which $Q(x,t \mid x',t')$ is the probability of an air parcel located at $x'$ at time $t'$ arriving at the receptor site $x$ at time $t$, $R(t \mid t')$ is the probability of the tracer not being reacted in the interval between $t'$ and $t$, $D(x',t')$ is the probability that the material is not lost by dry deposition at $(x',t')$ and $\Lambda(x',t')$ is the corresponding probability for not being lost through wet deposition. The integration is performed over a time period $\tau$, the length of the back trajectory calculation. The probabilities for deposition are poorly known and various formulations of these probability functions could be used. Keeler[22] assumes deposition to be linearly proportional to the concentration of the species of interest with the rate constants for dry ($k_d$) and wet ($k_w$) deposition to be defined as

$$k_d = \frac{v_d}{h} \tag{5}$$

$$K_h = \frac{\Theta(Z)\rho}{h} \tag{6}$$

with $v_d$ being the deposition velocity, $\Theta$ being the washout ratio for the species $Z$, $\rho$ being the precipitation rate ($\mathrm{mm\,hr^{-1}}$) and $h$ being the mixing height. If deposition velocity measurements were made as part of the studies, we can incorporate that information directly into this calculation, otherwise we would use estimates from the literature as done by Samson and Small.[26]

The transition probability function $Q(x,t\,|\,x',t')$ is approximated as a normal distribution about the trajectory center line with a standard deviation that increases linearly with time in the upwind direction[27,28] and can be expressed as

$$Q(x,t\,|\,x',t') = \frac{1}{2\pi\sigma_x(t')\sigma_y(t')}\exp\left[-\frac{1}{2}\left[\left[\frac{x''(t')}{\sigma_x(t')}\right]^2 + \left[\frac{y''(t')}{\sigma_x(t')}\right]^2 + \left[\frac{y''(t')}{\sigma_y(t')}\right]^2\right]\right] \tag{7}$$

where $x'' = X - x'(t')$ and $y'' = Y - y'(t')$ with $(X, Y)$ being the coordinates of the grid and $x'(t')$ and $y'(t')$ being the coordinates of the center line of the center line of the trajectory. The standard deviations of the trajectories are assumed to grow with time and can be approximated by

$$\sigma_x(t') = \sigma_y(t') = at' \tag{8}$$

with a dispersion speed, $a$, equal to $5.4\ \mathrm{km\,hr^{-1}}$.[20]

The integrated potential mass transfer for a given trajectory, $k$, arriving at time $t$ is then calculated as

$$\bar{T}_k(x\,|\,x') = \frac{\displaystyle\int_{t-\tau}^{t} T(x,t\,|\,x',t')\mathrm{d}t'}{\displaystyle\int_{t-\tau}^{t} \mathrm{d}t'} \tag{9}$$

This probability field predicts the possibility of contribution from any upwind area to a specific receptor site without regard to the actual location of the emission sources. These probabilities can then be incorporated into a Transport Bias Analysis by examining the ensemble average of potential mass transfer functions calculated for each trajectory over the same sampling period to obtain the mean potential mass transfer for that period. The spatial distribution of this field provides an estimate of the potential for contribution to the atmospheric

[26] P. J. Samson and M. J. Small, in *Modeling of Total Acid Precipitation Impacts*, ed. J. L. Schnoor, Butterworth, Boston, 1984, 1–24.
[27] R. R. Draxler and A. D. Taylor, *J. Appl. Meteorol.*, 1982, **21**, 367.
[28] P. J. Samson, *J. Appl. Meteorol.*, 1980, **19**, 1382.

concentration of a constituent if the source of that species were spatially homogeneous in the horizontal plane.

The measured concentrations of the species can then be used to calculate the transport bias. Using the potential mass transfer fields calculated for each trajectory, $k$, weighted by the corresponding concentration measured during that sampling interval results in a concentration-weighted mass transfer potential field as defined by

$$\bar{T}(x \mid x') = \frac{\sum_{k=1}^{K} \bar{T}_k(x \mid x') z_k(Z)}{\sum_{k=1}^{K} z_k(Z)} \tag{10}$$

where $\bar{T}_k$ is the integrated potential mass transfer calculated in eqn. (9) and $z_k(Z)$ is the concentration of the species of interest measured at $x$ when trajectory $k$ arrived. The difference between the weighted and the unweighted field represents the degree of bias associated with the transport of the species of interest to the receptor site.

One final manipulation is then made to convert the Transport Bias Analysis to Quantitative Transport Bias Analysis (QTBA). The concentration-weighted mass transfer potential of eqn. (10) is divided by the unweighted mass transfer potential, resulting in a dimensionless ratio. The ratio is then multiplied by the mean concentration of the species of interest, '$\bar{x}$' yielding

$$P(x \mid x') = \frac{\bar{T}(x \mid x')}{\bar{T}_k(x \mid x')} \bar{x} \tag{11}$$

This approach has been applied by Keeler and Samson[23] to the transport of particles to various locations in the northern US.

*Potential Source Contribution Function.* The Potential Source Contribution Function (PSCF) receptor model was originally developed by Ashbaugh *et al.*[20] and Malm *et al.*[29] It has been applied in a series of studies over a variety of geographical scales.[29–34] In a PSCF analysis, both chemical and meteorological data for each filter sample are needed. Air parcel back trajectories ending at a receptor site are calculated from the meteorological data with a trajectory model. Trajectories are represented by segment endpoints. Each endpoint has two coordinates (*e.g.* latitude, longitude) representing the central location of an air parcel at a particular time. To calculate the PSCF, the whole geographic region covered by the trajectories is divided into an array of grid cells whose size is

[29] W.C. Malm, C.W. Johnson and J.F. Bresch, in *Receptor Methods for Source Apportionment*, ed. T.G. Pace, Air Pollution Control Association, Pittsburgh, 1986, pp. 127–148.
[30] M.D. Cheng, P.K. Hopke and Y. Zeng, *J. Geophys. Res.*, 1993, **98**, 16839.
[31] N. Gao, M.D. Cheng and P.K. Hopke, *Anal. Chim. Acta*, 1993, **277**, 369.
[32] N. Gao, M.D. Cheng and P.K. Hopke, *Atmos. Environ.*, 1994, **28**, 1447.
[33] M.D. Cheng, N. Gao and P.K. Hopke, *J. Environ. Engineer.*, 1996, **122**, 183.
[34] N. Gao, P.K. Hopke and N.W. Reid, *J. Air Waste Manage. Assoc.*, 1996, **46**, 1035.

dependent on the geographical scale of the problem so that the PSCF will be a function of locations as defined by the cell indices *i* and *j*.

Let $N$ be the total number of trajectory segment endpoints during the whole study period, $T$. If $n$ segment trajectory endpoints fall into the $ij$th cell (represented by $n_{ij}$), the probability of this event, $A_{ij}$, is given by

$$P[A_{ij}] = \frac{n_{ij}}{N} \tag{12}$$

where $P[A_{ij}]$ is a measure of the residence time of a randomly selected air parcel in the $ij$th cell relative to the time period $T$.

Suppose in the same $ij$th cell there is a subset of $m_{ij}$ segment endpoints for which the corresponding trajectories arrive at a receptor site at the time when the measured concentrations are higher than a pre-specified criterion value. In this study, the criteria values were the calculated mean values for each species at each site. The probability of this high concentration event, $B_{ij}$ is given by $P[B_{ij}]$

$$P[B_{ij}] = \frac{m_{ij}}{N} \tag{13}$$

Like $P[A_{ij}]$, this subset probability is related to the residence time of air parcel in the $ij$th cell but the probability $B$ is for the contaminated air parcels.

The potential source contribution function (PSCF) is defined as

$$P_{ij} = \frac{P[B_{ij}]}{P[A_{ij}]} = \frac{m_{ij}}{n_{ij}} \tag{14}$$

$P_{ij}$ is the conditional probability that an air parcel which passed through the $ij$th cell had a high concentration upon arrival at the trajectory endpoint. Although the trajectory segment endpoints are subject to uncertainty, a sufficient number of endpoints should provide accurate estimates of the source locations if the location errors are random and not systematic.

Cells containing emission sources would be identified with conditional probabilities close to one of trajectories that have crossed the cells effectively transport the emitted contaminant to the receptor site. The PSCF model thus provides a means to map the source potentials of geographical areas. It does not apportion the contribution of the identified source area to the measured receptor data. In order to perform an apportionment, additional information must be included.

The elemental data determined for samples collected at Burbank, Claremont and Rubidoux during the 1987 Southern California Air Quality Study (SCAQS) were subjected to principal components analysis.[32] In those results it was found that the sulfur and nitrogen species were uncorrelated with the other elements representative of particle emission sources. Thus, PSCF was performed using air parcel backward trajectories for the intensive sampling dates calculated using the trajectory model developed by Russell and Cass.[35] Trajectories were calculated

[35] A. G. Russell and G. R. Cass, *Atmos. Environ.*, 1984, **18**, 1815.

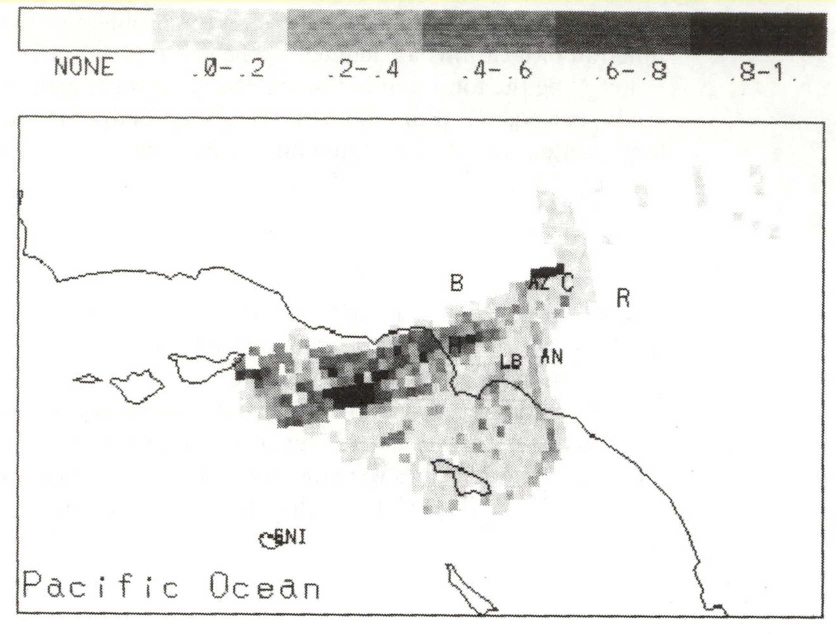


**Figure 5** PSCF map for $SO_2$ as measured at Claremont, CA, during SCAQS in summer 1987

on a terrain-following coordinate system. Each trajectory traces the air parcel movement backward in time (at an hourly interval) for one day at a height of 100 m above the terrain surface starting from the three receptor sites. This height (*i.e.* 100 m above the surface) was considered to be the most relevant for the pollutant transport in the SoCAB region. In this case, the grid cell dimensions were 5 km × 5 km. The PSCF maps for $SO_2$ and $SO_4^{2-}$ as measured at the Claremont site during the 55 periods of the summer intensive studies in the 1987 SCAQS project are given in Figures 5 and 6.

Several important differences in these maps can be noted. For the $SO_2$ map there is a group of high potential cells connecting the area around the Azuza sampling site (AZ) to C, the sampling site. This area does not appear as a high potential area in the $SO_4^{2-}$ map. There is an oil-fired power plant in that area where there would not be sufficient time in the air for the $SO_2$ to be converted into $SO_4^{2-}$. In contrast, in the $SO_4^{2-}$ map there is a high potential cell in the Long Beach area with a series of cells trailing from this cell into the ocean. This region does not appear in the $SO_2$ map. It is thought that this trailing effect arises because all of the trajectories that go from the off-shore area near Long Beach pass through the source region at Long Beach where there are a number of petroleum refineries. The emissions map for the SoCAB area for this time period is given in Figure 7. There is also a strong potential area off the coast that contributes to the $SO_2$, but not the $SO_4^{2-}$. It has been suggested[31] that this area is one in which sewage sludge has been deposited and thus there could be substantial emissions from dimethyl sulfide (DMS). The oxidation of DMS would lead to observed $SO_2$ at Claremont but not necessarily to increased $SO_4^{2-}$, whereas the $SO_2$ emitted in Long Beach would be expected to be converted to $SO_4^{2-}$ during the transport to Claremont.

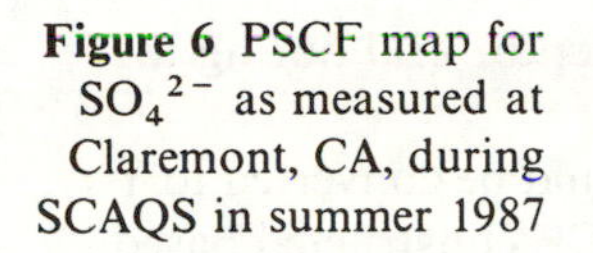

**Figure 6** PSCF map for $SO_4^{2-}$ as measured at Claremont, CA, during SCAQS in summer 1987

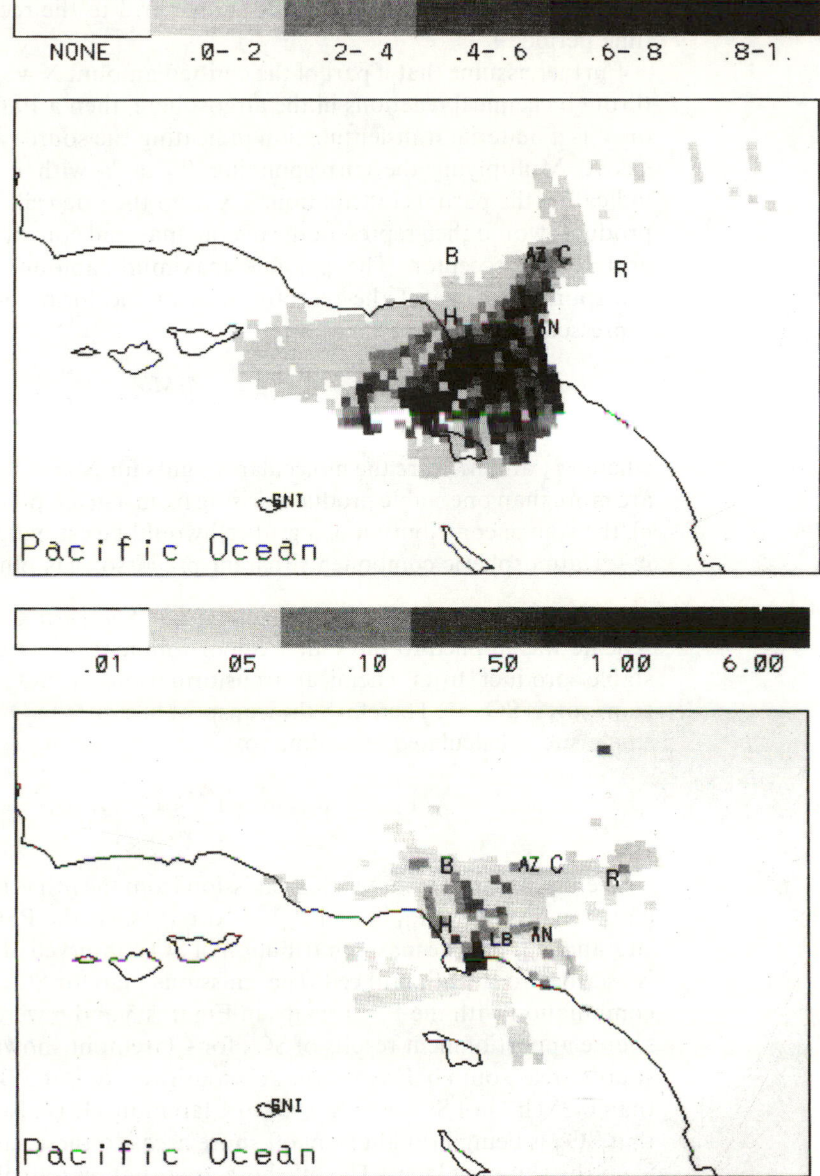


**Figure 7** Emissions estimates for the SoCAB as provided by A. G. Russell personal communication for August 25, 1987

*PSCF-based Source Apportionment Model.* To assign mass contribution to each grid in a map based on the information embedded in the PSCF values and emission data, a new model has to be devised. Consider the PSCF as an atmospheric material transfer function. Each grid PSCF value then indicates the likelihood that the observed concentrations at the receptor site are related to emissions from that source grid. Given a precursor gas X emitted at the grid cell with the emission rate $E$, the product of $P_{ij}$ and $E$ would then give a quantitative indication regarding the amount of X after atmospheric transformation and removal being transferred from the source grid to the receptor grid during the time period, $T$.

Further assume that if part of the emitted amount X would be converted to Y through chemical reactions in the atmosphere, then a PSCF constructed based on Y is a material transfer function indicating the source areas of its precursors gas X. Multiplying the corresponding $P_{ij}$ of Y with $E$ would yield a value indicating the partial contribution of X from the grid cell. The total of these two products would then represent the overall material contribution from the source grid to the receptor. The possible maximum amount of species, X, being transported to one of the receptor sites in the forms of X and $Y_n$ could be expressed as, $R_{ij}{}^{X}$ (t h$^{-1}$):

$$R_{ij}{}^{X} = P_{ij}{}^{X} E_{ij}{}^{X} + \sum_{1}^{n} \frac{M_{Y_n}}{M_X} P_{ij}{}^{Y_n} E_{ij}{}^{X} \tag{15}$$

where $M_X$ and $M_{Y_n}$ are the molecular weights for X and $Y_n$, respectively. If there are more than one stable products arising from a given precursor gas, $Y_i$ ($i = 1$ to $p$), the source contribution of a grid cell would be estimated as the sum of these $p + 1$ atmospheric components (parent precursor gas plus products).

*Source Apportionment for Sulfur Dioxide in Southern California.* The above scheme was applied to the source apportionment of $SO_2$. The most important stable product from chemical transformation of $SO_2$ during atmospheric transport is $SO_4{}^{2-}$. Therefore, the transport of emitted $SO_2$ in terms of a receptor single site is calculated according to:

$$R_{ij}{}^{SO_2} = P_{ij}{}^{SO_2} E_{ij}{}^{SO_2} + \left[\frac{M_{SO_4{}^{2-}}}{M_{SO_2}}\right] P_{ij}{}^{SO_4{}^{2-}} E_{ij}{}^{SO_2} \tag{16}$$

where $E_{ij}{}^{SO_2}$ is the estimated $SO_2$ emission from the $ij$th grid cell known from the emission inventory; $P_{ij}{}^{SO_2}$ and $P_{ij}{}^{SO_4{}^{2-}}$ are $P_{ij}$'s, *i.e.* the PSCF values for a single site; and $R_{ij}{}^{SO_2}$ is the mass contribution of $SO_2$ observed at a single receptor that is ascribed to the $ij$th grid cell. The emissions map for $SO_2$ shown in Figure 7 in combination with the PSCF maps in Figures 5 and 6 were used to produce the source apportionment results of $SO_2$ for Claremont shown in Figure 8. Several source areas south of Hawthorne and southwest of Long Beach contribute more than 0.25 t h$^{-1}$ of $SO_2$ and $SO_4{}^{2-}$ to Claremont. It is also seen in these figures that $SO_2$ is being brought from off-shore areas to these sampling sites, but only from the shipping lanes. The off-shore area that was attributed to the biogenic emission of DMS cannot be seen in Figure 8 because the area is not included in the emissions inventory. Dispersion models also are dependent on the completeness and accuracy of the emissions inventory so the inability to apportion $SO_2$ to the biogenic DMS source area would also not be included in a deterministic modeling effort.

*Source Apportionment for $NO_x$.* Cheng *et al.*[33] present a similar analysis for $NO_x$. The probability-weighted emissions map for $NO_x$ influencing $NO_y$ ($HNO_3$, $NO_3{}^-$, PAN) measure at Claremont is shown in Figure 9. Major sources ($Q > 1.7$ t h$^{-1}$) of $NO_x$ were identified by the emission inventory at the coastal

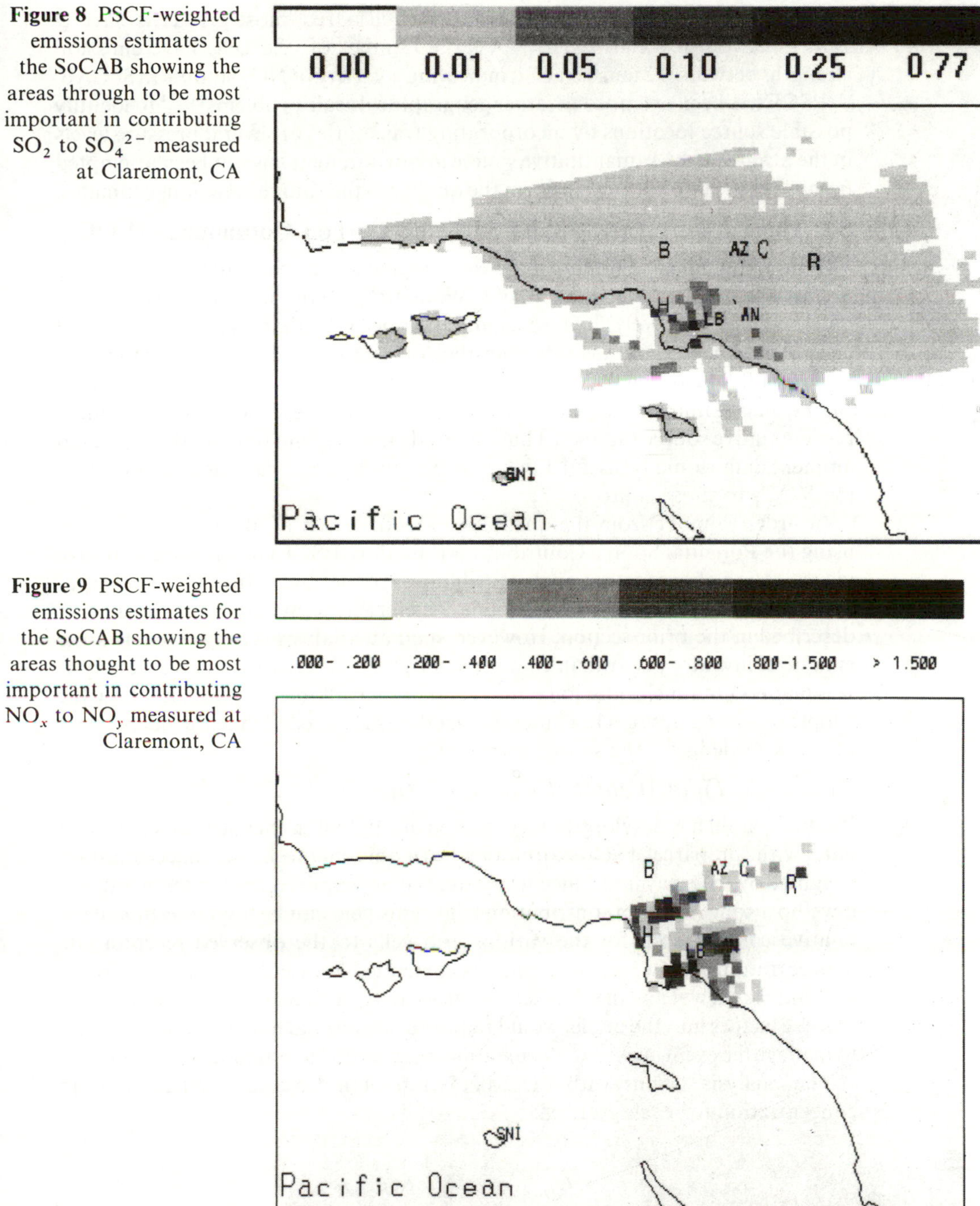


**Figure 8** PSCF-weighted emissions estimates for the SoCAB showing the areas through to be most important in contributing $SO_2$ to $SO_4^{2-}$ measured at Claremont, CA

**Figure 9** PSCF-weighted emissions estimates for the SoCAB showing the areas thought to be most important in contributing $NO_x$ to $NO_y$ measured at Claremont, CA

areas in two grid cells close to Long Beach (LB) and one cell close to Hawthorne (H). Several minor sources ($1.3 < Q \leq 1.7\ \mathrm{t\,h^{-1}}$) were located in the areas close to Long Beach and downtown Los Angeles areas. A few cells have $Q$ values between

0.9 and 1.3 t h$^{-1}$. In the downtown Los Angeles area, most grid cells have $Q$ values between 0.5 and 0.9 t h$^{-1}$. A large number of grid cells have emission strengths between 0.2 and 0.5 t h$^{-1}$, indicating a variety of $NO_x$ sources in SoCAB.

PSCF has been extended to larger geographical scale problems[30,34] to identify possible source locations by incorporating trajectories at several pressure levels in the atmosphere. A quantitative source apportionment has not been attempted because of the large uncertainties in the quality of the source emissions estimates.

## 5 Apportionment of Volatile Organic Carbon Compounds (VOCs)

VOC compounds can be apportioned in the same way as particulate matter using a CMB-like approach. Scheff and Wadden[36] have had good success in apportioning relatively low reactivity compounds that are generally well conserved during the transport from the source to the receptor site. However, there can still be problems with identification of the emission sources since there can be a large number of small sources and there can be difficulties in obtaining representative source samples. Thus, methods that depend only on the measured ambient data would be useful if they can identify source locations and apportion the VOCs to those sources.

In order to move from the qualitative identification of likely source regions using the Potential Source Contribution Function (PSCF) to a more quantative identification of relative regional contributions to the concentrations observed at a given site, PSCF was used to weight the emissions inventory for the area as described in the prior section. However, such an analysis is only possible if the emissions inventory is available on the same grid cell basis as the PSCF analysis was made. Often such emission estimates are not available and we cannot use the simple weighting approach. Thus, other methods are needed that are independent of our knowledge of the source inventories.

### *Residence Time Weighted Concentration*

Recently, Stohl has developed a new method of combining chemical concentration data with air parcel back trajectories.[37] In this method, the trajectories are weighted by the measured concentrations at one or more receptor sites in order to develop a concentration contribution field. This field can be used to estimate the relative contributions for the various grid cells to the observed receptor site concentrations. However, it cannot be extended to yield emission estimates without taking wet and dry deposition and dispersion into account. The inclusion of these factors into the model would require a substantial new effort, particularly to manage the event-by-event precipitation data needed to estimate wet deposition.

The analysis begins with the calculation of a trajectory-weighted mean concentration for each grid cell:

$$\overline{C_{mn}} = \frac{1}{\sum_l \tau_{mnl}} \sum_l \log(c_l)\tau_{mnl} \tag{17}$$

[36] P. A. Scheff and R. A. Wadden, in *Receptor Modeling for Air Quality Management*, ed. P. K. Hopke, Elsevier, Amsterdam, 1991, pp. 213–253.

[37] A. Stohl, *Atmos. Environ.*, 1996, **30**, 579.

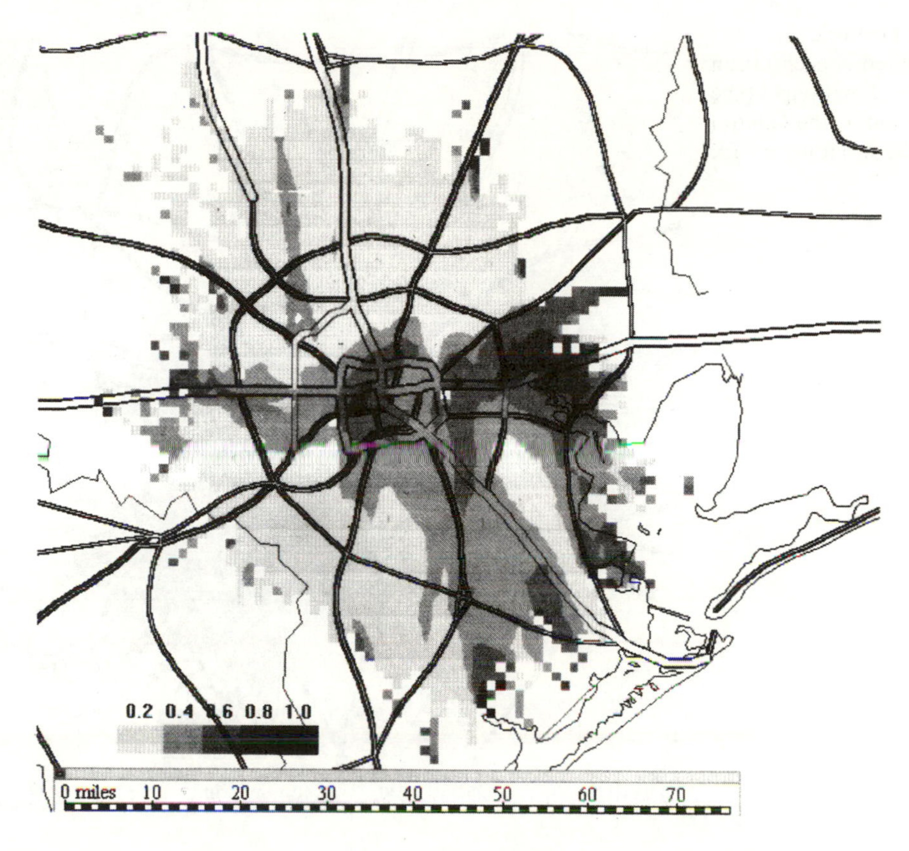


**Figure 10** Residence time weighted concentration for ethene measured at two sites in Houston, TX

where $m$ and $n$ are the indices of the grid cell, $l$ is the index of the trajectory segment, $c_l$ is the concentration at the receptor site associated with the arrival of trajectory $l$ (corrected for annual variation as discussed below), and $\tau_{mnl}$ is the time spent in the grid cell $(m,n)$ by trajectory $l$. A confidence interval is calculated for each cell. The concentration field is then smoothed with a nine-point filter which imposes the restriction that the values have to fall within their confidence band.

This approach weights each grid cell along the trajectory with the same value whereas the pollutants are input in specific locations. If two trajectories generally pass over the same cells except that one includes the pollutant emission grid cell, one will be associated with a low concentration while the other with a high value. Information from these two can be incorporated in the analysis as follows. Take $X_{il}$ as the mean concentrations of the grid cells which are hit by the segments $i = 1,\ldots,N$ of trajectory $l$ ($X_{il} = 10^{\overline{C_{mn}}}$ with $(m,n)$ being the respective grid cells), and $\overline{X_l}$ is the average concentration of the grid cells hit by the $N$ segments of trajectory $l$. The redistributed concentrations along trajectory $l$ are

$$c_{il} = c_l \frac{X_{il} N}{\sum_{j=l}^{N} X_{jl}} = c_l \frac{X_{il}}{\overline{X_l}} \qquad i = 1,\ldots,N \tag{18}$$

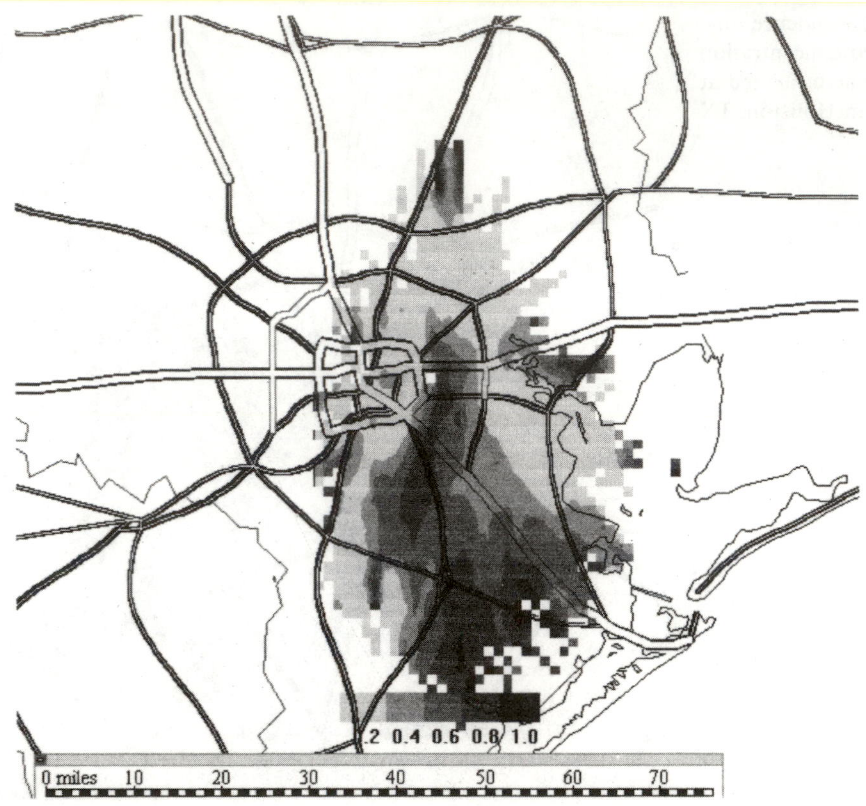


**Figure 11** Residence time weighted concentration for 2-methylpentane measured at the Clinton site in Houston, TX

where $c_l$ is the concentration associated with trajectory $l$. This redistributed field is smoothed and compared with the prior one. The process is repeated until there is less than 0.5% change from one iteration of the field to the next.

This method has been applied to speciated hydrocarbon data taken at two sites in Houston during the period of June 19 to November 30, 1993, using an automated gas chromatograph. The concentrations of 53 organic species were measured in 3900 samples. The trajectories were estimated from local surface meteorological data using the method of Dharmavaram and Hopke.[38] Figure 10 shows the results for ethene, a species emitted by motor vehicles, using the data from both sites. It can be seen that the areas of high concentration are focused around the highway network particularly to the east of the downtown ring road.

As an illustration of point source emissions, Figure 11 provides the residence time weighted concentration map for 2-methylpentane measured at a single site. Here it can be seen that several specific areas to the southeast of the downtown area have high contributions to the measured concentrations. At this time the contours represent the relative contributions of the area to the measured concentrations. Additional work is needed to develop this approach to the point where it will provide a quantitative apportionment of the source contributions.

[38] S. Dharmavaram and P. K. Hopke, *J. Air Pollut. Contr. Assoc.*, 1988, **38**, 812.

## 6 Summary

Receptor models, particularly the chemical mass balance model, have been effective in identifying sources of airborne particulate matter for development of control strategies for $PM_{10}$. With the imposition of stringent new ambient air quality standards, there will be additional requirements for identifying sources of gaseous emissions that lead to the formation of secondary particle mass or $O_3$. A number of new methods have been developed that offer the possibility to make such apportionments and thus effectively control the gaseous precursors that lead to secondary particles or $O_3$. These new methods need further development and testing, but their use should be considered by air quality managers as they develop their implementation plans.

[39] J. C. Chow, J. G. Watson, L. W. Richards, D. L. Haase, C. McDade, D. L. Dietrich, D. Moon and C. Sloane, *The 1989–90 Phoenix PM10 Study*, v. II: *Source Apportionment* [Final Report], report no. 8931.6F1, Arizona Department of Environmental Quality, Phoenix, AZ, 1991, table 3-1.

[40] W. M. Ryan, C. R. Badgett-West, D. R. Holtz, T. A. Peters, J. A. Cooper and D. Ono, *PM-10: Implementation of Standards*, APCA transactions series TR-13, Air Pollution Control Association, Pittsburgh, PA, 1988 pp. 419–429.

[41] L. C. Thanukos, T. Miller, C. V. Mathai, D. Reinholt and J. Bennett, in *$PM_{10}$ Standards and Nontraditional Particulate Source Controls*, v. I, J. C. Chow and D. M. Ono, eds., A&WMA transactions series no. 22, Air & Waste Management Association, Pittsburgh, PA, 1992, pp. 244–261.

[42] H. A. Gray, B. Landry, C. S. Liu, R. C. Henry, J. A. Cooper, and J. R. Sherman, in *PM-10: Implementaion of Standards*, C. V. Mathai, and D. H. Stonefield, eds., APCA transactions series TR-13, Air Pollution Control Association, Pittsburgh, PA, pp. 399–418.

[43] J. G. Watson, J. C. Chow, Z. Lu, E. M. Fujita, D. H. Lowenthal, D. R. Lawson, L. L. Ashbaugh, *Aerosol Sci. Technol.*, 1994, **21**, 1.

[44] K. L. Magliano, 'Level 1 $PM_{10}$ assessment in a California air basin', in *PM-10: Implementation of Standards*, C. V. Mathai, and D. H. Stonefield, eds., APCA transactions series TR-13, Air Pollution Control Association, Pittsburgh, PA, 1988, pp. 508–517.

[45] J. C. Chow, J. G. Watson, D. H. Lowenthal, P. A. Solomon, K. L. Magliano, S. D. Ziman, and L. W. Richards, *Atmos. Environ.*, 1992 **26A**, 3335.

[46] J. Cooper, J. Sherman, E. Miller, D. Redline, L. Valfovinos, and W. Pollard, in *PM-10: Implementation of Standards*, In: *PM-10: Implementation of Standards*, C. V. Mathai, and D. H. Stonefield, eds., APCA transactions series TR-13, Air Pollution Control Association, Pittsburgh, PA, 1988, pp. 430–437.

[47] B. M. Kim, M. D. Zeldin, and C. S. Liu, in *$PM_{10}$ Standards and Nontraditional Particulate Source Controls*, v. II, J. C. Chow and D. M. Ono, eds., A&WMA transactions series no. 22, Air & Waste Management Association, Pittsburgh, PA, 1992, pp. 979–991.

[48] J. C. Chow, C. S. Liu, J. Cassmassi, J. G. Watson, Z. Lu, and L. C. Pritchett, *Atmos. Environ.*, 1992, **26A**, 693.

[49] A. L. Dresser and B. K. Baird in *PM-10: Implementaion of Standards*, C.V. Mathai, and D. H. Stonefield, eds., APCA transactions series TR-13, Air Pollution Control Association, Pittsburgh, PA, 1988, pp. 458–469.

[50] J. E. Houck, J. A. Rau, S. Body, and J. C. Chow, in *$PM_{10}$ Standards and Nontraditional Particulate Source Controls*, v. II, J. C. Chow and D. M. Ono, eds., A&WMA transactions series no. 22, Air & Waste Management Association, Pittsburgh, PA, 1992, pp. 219–230.

[51] J. C. Chow, J. G. Watson, R. T. Egami, C. A. Frazier, Z. Lu, A. Goodrich, and A. Bird, *J. Air Waste Manage. Assoc.*, 1990, **40**, 1134.

[52] L. W. Skidmore, J. C. Chow, and T. T. Tucker, '$PM_{10}$ air quality assessment for the Jefferson County, Ohio air quality control region', *$PM_{10}$ Standards and Nontraditional Particulate Source Controls*, v. I, J. C. Chow and D. M. Ono, eds., A&WMA transactions series no. 22, Air & Waste Management Association, Pittsburgh, PA, 1992, pp. 1016–1031.

# The Critical Load Approach to Air Pollution Control

MICHAEL HORNUNG, HELEN DYKE, JANE R. HALL and SARAH E. METCALFE

## 1 Background

During the 1970s and 1980s a consensus gradually developed about the link between the deposition of atmospheric pollutants, primarily compounds of S and N, and the acidification of soils and freshwaters and of forest damage. The accumulation of knowledge on the link between pollutant emissions and environmental damage in the 1970s led to the negotiation of the Convention on Long Range Air Pollution (LRTAP), signed in 1979 under the auspices of the United Nations Economic Commission for Europe (UNECE). A number of protocols to reduce emissions have been signed under the convention: two on sulfur dioxide ($SO_2$) in 1985 and 1994, one on nitrogen oxides ($NO_x$) in 1988 and one on volatile organic compounds (VOCs) in 1991. The earliest agreements aimed at uniform percentage reductions; thus the first sulfur protocol proposed a 30% reduction in sulfur emissions or transboundary fluxes of sulfur compounds from the 1980 levels. Uniform, across the board reductions have a number of weaknesses. The variation in sensitivity of receptors, be they natural ecosystems or man-made materials, is not taken into account. A uniform reduction may not provide protection for specific valued receptors. Uniform reductions also take no account of the spatial relationship between the emission sources and the distribution of receptors.

The critical load and levels concept provides a receptor-based approach which can also incorporate variations in spatial patterns of emissions and impacts, and allows targeting of emission reductions and protection. The concept was first outlined in North America in the late 1980s during discussions between the USA and Canada on the control of transboundary pollution. It was developed into a workable approach to the assessment of pollution impacts, and hence setting of emission targets, through a series of workshops in Europe which were sponsored variously by the UNECE and/or national governments. These workshops have refined the concept and formalized methodologies for the calculation and mapping of critical loads and critical levels and current recommended methodologies are now published as a handbook.[1,2]

[1] *Calculation and Mapping of Critical Thresholds in Europe: Status Report 1995*, ed. M. Posch, P. A. M de Smet, J.-P. Hetteling and J. P. Downing, RIVM, Bilthoven, 1995.

[2] Federal Environment Agency, *Manual on Methods for Mapping Critical Loads/levels and Geographical Areas where they are Exceeded*, Federal Environment Agency, Berlin, 1996.

The early focus of critical load mapping was the impacts of deposited S and acidity. During the late 1990s the focus has moved to nitrogen and work is now in progress on approaches for heavy metals and persistent organic pollutants (POPs). The Second Sulfur Protocol, in 1994, took account of the critical loads approach and the current negotiations of the second nitrogen protocol are drawing upon critical loads information. The original context was the control of transboundary air pollution but more recently the concept has also been applied at the scale of the specific site, catchment or single power plant. The present review considers the methods which have been developed to calculate and map critical loads for acidic pollutants. The examples are drawn from the UK but the methodologies are those being used throughout Europe.

## 2 Definitions

The most commonly used general definition of critical load is:

> 'A quantitative estimate of an exposure to one or more pollutants below which significant harmful effects on specified sensitive elements of the environment do not occur according to present knowledge'.[3]

Pollutant specific definitions have also been developed, for example:

> (i) for acidity—the highest deposition of acidifying compounds that will not cause chemical changes leading to long term harmful effects on ecosystem structure and function'.[3]
> (ii) for nutritional effects of nitrogen—a quantitative estimate of an exposure to deposition of nitrogen as $NH_x$ and/or $NO_y$ below which harmful effects in ecosystem structure and/or function do not occur according to present knowledge.[1]

The definitions and the approach are implicitly based on a dose–response relationship between pollutant and receptor, or indicator organism. In principle, the critical load would be set at the point at which a measurable response is detected. A formal dose–response relationship is, however, rarely available for a given pollutant–receptor combination.

## 3 Calculation of Critical Loads

A number of approaches have been developed for the calculation and mapping of critical loads. These are generally grouped as: level 0 or empirical, level 1 or mass balance based approaches and level 2 or dynamic modelling approaches. The choice of method is usually determined by the availability of input data and the application.

[3] *Critical Loads for Sulphur and Nitrogen*, ed. J. Nilsson and P. Grennfelt, Nordic Council of Ministers, Copenhagen, 1988.

## *Empirical Approaches*

The empirical approaches range from a ranking of ecosystems in terms of their sensitivity to a given pollutant, to assigning critical loads on the basis of data on properties of the system which control the response to a given pollutant and empirically derived models based on, for example, the results of surveys of water chemistry across pollution gradients.

*Critical Loads of Acidity for Soils.* Mineral weathering in soils provides the main long-term sink for deposited acidity and an early approach set the critical load for acidity as the amount of acidity which could be buffered by the annual production of base cations from mineral weathering.[3] An empirical approach based on this principle has been developed in the UK and used to produce national critical load maps.

Soils were assigned to a critical load class based on their content of different types of weatherable minerals, for example carbonates and the various types of layered silicates, and on soil chemistry.[4] Maps were produced using the databases of the Soil Survey and Land Use Research Centre (SSLRC) and the Macaulay Land Use Research Institute (MLURI). A critical load was assigned to each 1 km square of the UK National Grid on the basis of the mineralogy and chemistry of the dominant soil occurring in the grid cell. Peat soils contain very little mineral material and a different approach was used where peat was the dominant soil. The critical load was set using an approach[5] which assumes that peat acidity comes to an equilibrium with the incoming rainfall acidity. The critical load is calculated as that input of deposited acidity which would result in a decrease in peat pH of no more than 0.2 pH units compared to the pH which would obtain under pristine conditions. A critical load was therefore assigned to each 1 km square of the National Grid. As applied in the UK, this represents a precautionary approach as it will effectively set the critical load for mineral soils to prevent any further change in soil chemistry as a result of deposited acidity.

*Critical Loads of S and Acidity for Surface Waters: The Steady State Water Chemistry Model (SSWC).* The SSWC approach was developed in Norway[6] and has been used to calculate and map critical loads in a number of European countries, including the UK, and North America. It is a steady-state approach which assumes that inputs and outputs are in balance, enabling mean values for water chemistry and deposition to be used as input data. It calculates the critical load to ensure that (i) base cation production within the catchment, from mineral weathering, is greater than acidic inputs and (ii) that a target value of acid neutralizing capacity (ANC) is maintained within the lake or stream; ANC is defined as the difference between base cations and strong acid anions in solution.[7]

[4] M. Hornung, K. R. Bull, M. Cresser, J. Hall, S. J. Langan, P. Loveland and C. Smith, *Environ. Pollut.*, 1995, **90**, 301.

[5] C. S. S. Smith, M. S. Cresser and R. D. J. Mitchell, *Ambio*, 1993, **22,** 22.

[6] A. Hendriksen, W. Dickson and D. F. Brakke, in *Critical Loads for Nitrogen and Sulphur*, ed. J. Nilsson and P. Grennfelt, Nordic Council of Ministers, Copenhagen, 1988, p. 87.

[7] J. O. Reuss and D. W. Johnson, *Acid Deposition and the Acidification of Soils and Waters*, Springer, New York, 1986.

The target ANC value is set to ensure the survival of populations of an indicator organism(s). The concentrations of the main chemical determinands in waters which influence the distribution and survival of freshwater aquatic biota, *e.g.* pH, $Ca^{2+}$, $Al^{3+}$, are highly correlated with ANC. In the UK the target ANC used in the calculation of critical loads is 0 $\mu eq\,l^{-1}$, which indicates a 50% probability of brown trout survival.[8] A value of 20 $\mu eq\,l^{-1}$ is generally used in Scandinavia. The probabilities of survival are obtained by the application of logistic regression to data from surveys of water quality and fishery status, or fish survival studies.

The critical load is calculated using:

$$CL = ([BC]_0{}^* - [ANC_{limit}])Q - [BC]_d{}^*R \tag{1}$$

where CL = critical load of acidity, * = indicates a non-marine component, $[BC]_0$ = original sea salt corrected base cation concentration prior to acidification, $ANC_{limit}$ = critical level appropriate to a target organism, $Q$ = runoff, $[BC]_d$ = non-marine base cation concentration in precipitation and $R$ = rainfall.

$BC_0$ is calculated from the following equation as data are not available:

$$[BC]_0{}^* = [BC]_t{}^* - F([SO_4]_t{}^* - [SO_4]_0{}^*) \tag{2}$$

where $[SO_4]_t{}^*$ = present day excess sulfate and $[BC]_t$ = present day excess base cation concentration. The value for the '*F* factor' is usually calculated from an equation derived empirically[9] using the results from a large survey of lake water chemistry across pollution gradients in Norway.

The methodology has now been extended to allow critical loads of sulfur and acidity to be considered simultaneously using a mass balance approach, the First Order Acidity Balance (FAB) Model.[1,2]

*Critical Loads of S Acidity for Surface Waters: The Empirical Diatom Model.* This model has been developed from palaeolimnological data and diatom-based pH reconstructions. The diatom assemblage present in a lake is sensitive to pH. Diatom records from acidified lakes generally show that the diatom flora was relatively stable for long periods prior to acidification, indicating relatively stable water chemistry. Acidification is indicated by a shift to a more acid-tolerant diatom flora. The approach assumes that the point of change towards the more acidophilous diatom flora indicates the time at which the critical load was exceeded.[10,11] As it is based on the point of onset of acidification, the approach is said to set a 'base critical load'.

The method assumes that acidification is a function of the sensitivity of the lake

[8] S. Juggins, S. J. Ormerod and R. Harriman, in *Critical Loads of Acid Deposition for United Kingdom Freshwaters*, Institute of Terrestrial Ecology, Edinburgh, 1995, p. 9.

[9] L. Lein, G. G. Raddum and A. Fjellheim, *Critical Loads of Acidity to Freshwater—Fish and Invertebrates*, Report 0-89185, Norwegian Institute for Water Research, Oslo, 1992.

[10] R. W. Battarbee, T. E. H. Allott, A. M. Krieser and S. Juggins, in *Critical Loads: Concepts and Applications*, ed. M. Hornung and R. A. Skeffington, HMSO, London, 1993, p. 99.

[11] R. Battarbee, T. E. H. Allott, S. Juggins and A. M Kreiser, in *Critical Loads of Acid Deposition for United Kingdom Freshwaters*, Institute of Terrestrial Ecology, Edinburgh, 1995, p. 3.

to acid deposition. Calcium in solution is used as a measure of sensitivity and S deposition as driving acidification. Data from a series of lake sites in the UK were used to identify the ratio of lake water $Ca^{2+}$ to S load in current deposition, which most clearly separated the acidified from the non-acidified lakes. The optimal ratio, determined by logistic regression, was 94:1 with acidified sites having a lower ratio.

The critical load can be defined as the acidic deposition at a site for the 94:1 ratio. However, the calcium concentration will have changed since acidification and so a precautionary approach is taken by dividing $Ca_0$, the pre-acidification calcium concentration, by the critical Ca:S ratio. $Ca_0$ is calculated from current mean calcium values using the empirically derived '*F* factor' discussed above. The resultant is re-expressed as keq $ha^{-1}$ $yr^{-1}$.

The model has been recalibrated to allow the calculation of a critical load for total acidity, including the effects of nitrogen deposition.[12] The critical ratio for total acidity was determined as 89:1. The pre-acidification levels are calculated as:

$$[Ca^{2+}]_0 = [Ca^{2+}]_t - F_{Ca_0}([SO_4{}^{2-}]_t{}^* + [NO_3{}^-]_t - [SO_4{}^{2-}] - [NO_3{}^-]_0) \quad (3)$$

where $_0$ indicates pre-industrial values, $_t$ indicates current day values, * indicates non-marine components and $F_{Ca_0}$ is the factor for calculating pre-industrial calcium.

*Application of the Empirical Approaches for Surface Waters in Mapping Critical Loads in the UK.* Both the above approaches have been applied in mapping critical loads in the UK. The input water chemistry data were derived from a national survey and data on deposition of S or acidity from available national datasets. The national water sampling programme sampled what was thought to be the most sensitive water body in each cell of a 10 km grid in those areas of the UK with high or moderate sensitivity to acidification (as judged on the basis of soils and geology), and in a 20 km grid in areas which had low sensitivity or were considered insensitive to acidic inputs.[13] Within the grid cells, the water bodies sampled were selected on the basis of published schemes for predicting groundwater sensitivity based on geology[14] or surface water sensitivity based on data on soils and geology.[15] If these schemes identified a number of waterbodies with similar sensitivities, the water body at the highest altitude was sampled. A minimum size for sampled waterbodies was set at 0.5 ha and large waterbodies which extended between grid squares were excluded. The available deposition data is on a 20 km grid and the critical load for a given waterbody was calculated using the deposition data from the relevant 20 km grid cell. The calculated critical

[12] T. E. H. Allott, R. W. Battarbee, C. Curtis, R. Harriman, J. Hall, K. R. Bull and S. E.Metcalfe, in *Critical Loads of Acid Deposition for United Kingdom Freshwaters*, Institute of Terrestrial Ecology, Edinburgh, 1995, p. 25.

[13] A. M. Krieser, S. T. Patrick, R. W. Battarbee, J. Hall and R. Harriman, in *Critical Loads of Acid Deposition for United Kingdom Freshwaters*, Institute of Terrestrial Ecology, Edinburgh, 1995, p. 15.

[14] W. M. Edmunds and D. G Kinneburgh, *J. Geol. Soc.*, 1986, 707.

[15] M. Hornung, K. R. Bull, M. Cresser, J. Ullyet, J. R. Hall, S. J. Langan, P. J. Loveland and M. J. Wilson, *Environ. Pollut.*, 1995, **87**, 99.

**Table 1** Summary of empirical critical loads for nitrogen deposition to (semi-)natural freshwater and terrestrial ecosystems

| | *Critical load/* kg N $ha^{-1}$ $yr^{-1}$ | *Indication of exceedance* |
|---|---|---|
| **Trees and forest ecosystems** | | |
| Coniferous trees (acidic) (low nitrification rate) | 10–15## | Nutrient imbalance |
| Coniferous trees (acidic) (mod.–high nitrification rate) | 20–30## | Nutrient imbalance |
| Deciduous trees | 15–20## | Nutrient imbalance; increased shoot/root ratio |
| Acidic coniferous forests | 7–20## | Changes in ground flora and mycorrhizae; increased leaching |
| Acidic deciduous forests | 10–20# | Changes in ground flora and mycorrhizae |
| Calcareous forests | 15–20(#) | Changes in ground flora |
| Acidic forests* | 7–15(#) | Changes in ground flora and leaching |
| Forests in humid climates | 5–10(#) | Decline in lichens and increase in free-living algae |
| **Heathlands** | | |
| Lowland dry heathlands | 15–20## | Transition heather to grass; functional change (litter production; flowering; N accumulation) |
| Lowland wet heathlands | 17–22# | Transition heather to grass |
| Species-rich heaths/acid grassl. | 10–15# | Decline in sensitive species |
| Upland Calluna heaths | 10–20(#) | Decline in heather dominance, mosses and lichens; N accumulation |
| Arctic and alpine heaths* | 10–20(#) | Decline in lichens, mosses and evergreen dwarf shrubs |
| **Species-rich grasslands** | | |
| Calcareous grasslands | 15–35# | Increased mineralization, N accumulation and leaching; increase in tall grass, change in diversity† |
| Neutral–acid grasslands | 20–30# | Increase in tall grass, change in diversity |
| Montane–subalpine grasslands | 10–15(#) | Increase in tall graminoids, change in diversity |
| **Wetlands** | | |
| Mesotrophic fens | 20–35# | Increase in tall graminoids, decline in diversity |
| Ombrotrophic bogs | 5–10## | Decrease typical mosses, increase in tall graminoids, N accumulation |
| Shallow soft-water bodies* | 5–10## | Decline in isoetid species |

## Reliable; #quite reliable and (#) expert judgement.
* = Unmanaged, natural systems. † = use low end of the range for N limited, high end for P limited calcareous ecosystems.

load was used to allocate the grid cell in which the waterbody occurred to a critical load class.

The resultant maps from the two approaches show very similar patterns, with areas with small critical loads in the uplands of the north and west but also in a few parts of the south and east of England underlain by base-poor sands. However, more grid cells are in the classes with the smallest critical loads on the map derived using the diatom model, *i.e.* the diatom model tends to give lower critical loads than the SSWC.

*Critical Loads of Nutrient Nitrogen.* Enhanced N deposition to terrestrial and freshwater ecosystems can lead to acidification or eutrophication. The latter can have major impacts on plant competition within vegetation habitats, leading to changes in species composition, and on the sensitivity of vegetation to environmental stresses, such as drought or frost, and insect predation.

Critical loads for N in the context of eutrophication have been set for a number of woodland, grassland, heathland and bog communities using the results from experimental studies and field observations, or 'expert judgement'.[2,16] The critical load for a given habitat is expressed as a range of values to indicate the uncertainty and to allow for real intra-habitat variations between regions (Table 1). The reliability of the assigned critical loads varies from 'reliable', where it is based on several publications which show concordant results, to 'expert judgement', where no published data are available at present.

In the UK, critical load maps have been created by assigning the relevant critical load to distribution maps of the habitats. The Institute of Terrestrial Ecology's Land Cover Map of Great Britain[17] or a combination of the National Vegetation Classification[18] and data from the Biological Records Centre[19] were used to define the distribution of the habitats.[20] The relevant critical load was assigned to each 10 km square of the National Grid in which the habitat was predicted to occur. For example, this approach has been used to produce critical load maps for heathland, sub-alpine grassland and coniferous forest.[20]

## Mass Balance Based Approaches

These methods are primarily used to calculate long-term critical loads for systems at steady state. Factors which can affect the impact of a pollutant in the short term, such as sulfate adsorption, are therefore not included. A mass balance based approach is currently the most widely used method in Europe for producing critical load maps for acidity, S and N. Mass balance based approaches are currently under development for heavy metals and persistent organic pollutants.

[16] R. Bobbink, M. Hornung and J. G. M. Roelofs, in *Manual for Mapping Critical Loads/Levels and Geographical Areas where they are Exceeded*, Federal Environment Agency, Berlin, 1996, p. III-1.

[17] R. Fuller and G. Groom, *GIS Europe*, 1993, **2**, 25.

[18] *British Plant Communities. Vol. 3: Grasslands and Montane Communities*, ed. J. S. Rodwell, Cambridge University Press, Cambridge, 1993.

[19] P. T. Harding and J. Sheail, in *Biological Recording of Changes in British Wildlife*, ITE Symposium no. 26, ed. P. T. Harding, HMSO, London, 1992, p. 5.

[20] K. R. Bull, M. J. Brown, H. Dyke, B. C. Eversham, R. M. Fuller, M. Hornung, D. C. Howard, J. Rodwell and D. B. Roy, *Water, Air, Soil Pollut.*, 1995, **85**, 205.

The simplest approaches solve the mass balance model for a single layer soil, but computer models have been developed which can be applied to a multi-layered soil, for example the PROFILE[21] and MACAL models.[22]

*Mass Balance Approach for Nutrient Nitrogen.* The approach is based on an equation which aims to balance all significant long-term inputs and outputs of N for terrestrial systems. In this context, long term is defined as at least one forest rotation or 100 years. The critical load is set (i) to prevent an increase in leaching of nitrogen compounds, particularly nitrate which will result in damage to the terrestrial or linked aquatic ecosystems, and (ii) to ensure sustainable production by limiting N uptake and removal to a level which will not result in deficiencies of other nutrient elements. The equation is currently formulated as follows:

$$\mathrm{CL\ nut(N)} = N_i + N_u + N_{de} + N_{fire} + N_{eros} + N_{vol} + N_{le} - N_{fix} \quad (4)$$

where CL nut(N) = critical load of nutrient nitrogen, $N_{fix}$ = N inputs from biological fixation of N, $N_i$ = acceptable level of N immobilization in soil organic matter, $N_u$ = net removal of N in harvested vegetation and/or animals, $N_{de}$ = annual flux to the atmosphere of N by denitrification, $N_{fire}$ = losses of N in smoke from natural and wildfires or fires used as part of traditional management, $N_{eros}$ = losses of N through erosion under natural conditions and following forest harvesting or fire, $N_{vol}$ = N losses to the atmosphere through volatilization of ammonia and $N_{le}$ = an acceptable level of N leaching ($NO_3$ + $NH_4$ + organic N) from the rooting zone which will not result in damage in the terrestrial or linked aquatic ecosystems plus any enhanced leaching following forest harvest or fires. All values are expressed in eq $ha^{-1}$ $yr^{-1}$. Immobilization of ammonium on mineral surfaces is not included as this is considered a short-term sink.

In principle the approach could be used for any terrestrial ecosystem or vegetation habitat, but to date its use has been largely restricted to forest systems. All input values to the equation are expressed in eq $ha^{-1}$ $yr^{-1}$, equivalent to $mol_c$ $ha^{-1}$ $yr^{-1}$. Thus, for example, removal of N in forest harvest and enhanced leaching after harvest are divided by the length of the forest rotation to derive the long-term average rate of removal/loss. Measured values for all the above variables are rarely available for the construction of national datasets and the development of national critical load maps. As a result, default values derived from surveys of published values are often used.[2] In many ecosystems a number of the variables can be omitted from the equation as the losses are insignificant, for example $N_{fire}$, $N_{eros}$ and $N_{vol}$.

In the UK the equation has been used to calculate a critical load for forests. The equation was solved with respect to a Sitka spruce crop of mean national yield class, or oak of mean average yield class growing on the dominant soil in each 1 km square of the National Grid. Data on the yield class and on wood chemistry

[21] P. Warfvinge and H. Sverdrup, *Water, Air, Soil Pollut.*, 1992, **63**, 119.

[22] W. de Vries, *Methodologies for the Assessment and Mapping of Critical Loads and the Impact of Abatement Strategies on Forest Soils*, DLO Winand Staring Centre for Integrated Land, Soil and Water Research, Bilthoven, 1991.

allowed the N removed at harvest to be calculated. Default values were used for $N_{de}$ and $N_i$ with the value varied in relationship to soil moisture class and pH, respectively. A uniform value of $N_{le}$ was used, derived from available data on leaching from forests in relatively unpolluted areas and on leaching losses after felling. The resultant critical load value was assigned to the whole 1 km square.

*Critical Loads of Acidity for Forest Soils.* The approach is based on balancing inputs of acidity to the ecosystem with sinks within the system and outputs. The critical load is calculated such that the balance between inputs, sinks and outputs ensures that a chosen critical chemical limit is not exceeded. The critical chemical limit is set to protect a biological indicator which is used as an indicator of the 'health' of the forest. The most widely used critical chemical limit is the ratio of base cations, or calcium to aluminium, in soil solution at which fine root damage occurs; a value of 1 is commonly used for the ratio. Critical chemical limits set in terms of pH or the concentration of aluminium in soil solution have also been proposed.

The early methodologies calculated a critical load for acidity. However, negotiations of emission reduction targets, or setting of emission limits, are in terms of sulfur or nitrogen compounds rather than acidity. The approach has therefore been modified in recent years to allow the calculation of critical loads of S plus N. This allows the two pollutants to be treated separately but also allows interactions to be considered and the relative impacts of controlling each of the pollutants.

The equation which forms the basis for the calculation of the critical loads has been derived from a charge balance of ions in leaching fluxes from the soil compartment combined with mass balance equations for the inputs, sinks, sources and outputs of sulfur and nitrogen.[1] The critical load of actual acidity is expressed as:

$$CL(acact) = BC_w - ANC_{le(crit)} \tag{5}$$

and the critical load of S plus the critical load of N as[1,2]

$$CL(S) + CL(N) = BC_{dep} - Cl_{dep} + BC_w - BC_u + N_i + N_u + N_{de} - ANC_{le(crit)} \tag{6}$$

where $CL(acact)$ = critical load of actual acidity, $CL(S)$ = critical load of S, $CL(N)$ = critical load of N, $Cl_{dep}$ = deposition of chloride, $BC_{dep}$ = deposition of base cations, $BC_w$ = base cation production from mineral weathering, $BC_u$ = net base cation removal in harvested timber, $N_i$ = nitrogen immobilization in soil organic matter, $N_u$ = N removal in harvested timber, $N_{de}$ = N losses by denitrification and $ANC_{le(crit)}$ = a critical level of leaching of alkalinity which will not result in 'harmful' effects to the ecosystem. All values are expressed in eq $ha^{-1}$ $yr^{-1}$.

The $ANC_{le(crit)}$ is calculated as the permitted level of leaching of H and Al in solution. If the critical chemical limit is set in terms of the base cation to aluminium ratio in soil solution at which root damage occurs, the term for $ANC_{le(crit)}$ is converted to:[1,2]

$$1.5 \times \frac{BC_{dep} + BC_w - BC_u}{(BC/Al)_{crit}} + Q^{2/3} \times 1.5 \times \frac{BC_{dep} + BC_w - BC_u}{(BC/AL)_{crit} \times K_{gibb}} \quad (7)$$

where $Q$ = flow of water from the soil, in $m^3\,ha^{-1}\,yr^{-1}$, $(BC/AL)_{crit}$ = critical base cation to aluminum ratio in soil solution and $K_{gibb}$ = gibbsite equilibrium constant. The multipliers 1.5 and 2/3 arise from conversions from mols to equivalents.

In the UK, the mass balance equation has been used to calculate critical loads of acidity for deciduous and coniferous forest. The equation was solved for each 1 km square of the National Grid. For coniferous forest, the calculations were based on Sitka spruce of a mean national yield class growing on the dominant soil type in each 1 km grid cell and for deciduous woodland on an oak crop of a mean national yield growing on the dominant soil in each cell. $BC_w$ was derived for the dominant soil in each 1 km grid cell using information on mineralogy and soil chemistry. $BC_u$ was derived from the yield class and wood chemistry for Sitka spruce or oak. Runoff was derived from a national data set on the 1 km grid basis. The BC:Al ratio was set to a value of 1. The resultant critical load maps were then overlain with distribution maps for coniferous woodland or deciduous woodland, derived from the ITE Land Cover Map of Great Britain, to derive a critical load map for the stock at risk.

## *The Critical Load Function: Assessment of Pollutant Interactions*

The deposition of sulfur and nitrogen can be related to the critical load of acidity using the critical load function (Figure 1).[1,23] This shows the critical load for all combinations of sulfur and nitrogen deposition. However, if the deposition of S *versus* N is compared to the critical load, the nitrogen sinks ($N_i$, $N_u$, $N_{de}$) cannot compensate for deposited sulfur deposition. The maximum critical load for sulfur, $CL_{max}(S)$ (Figure 1) is therefore given as:

$$CL_{max}(S) = BC_{dep} - Cl_{dep} + BC_w - BC_u - ANC_{le(crit)} \quad (8)$$

Further, if N deposition is equal to or less than the sink terms $N_i$, $N_u$ and $N_{de}$, all deposited N is consumed by the sinks and sulfur can be considered alone; the value for $N_{dep} < N_i + N_u + N_{de}$ is referred to as $CL_{min}(N)$.

Consideration of the critical load of nutrient N can be combined with the acidity critical loads. If the critical load of nutrient N is less than $CL_{max}(N)$, then $CL_{nut}(N)$ becomes the maximum permitted N deposition.

If actual or modelled S and N deposition data are plotted onto Figure 1, the function can be used to examine options for the reduction of S or N deposition, and hence emissions, to protect a given receptor at the site or grid cell scale.

## *Dynamic Models*

The dynamic models incorporate processes which control the sources and sinks of acidity within the system, the rate of production and consumption of alkalinity,

[23] K. R. Bull, *Water, Air, Soil Pollut.*, 1995, **85**, 201.

1 only sulfur reduction will offer protection
2 sulfur reductions may subsequently offer the option of reducing sulfur or nitrogen
3 sulfur and nitrogen must be reduced before there is any option of reducing either sulfur or nitrogen
4 nitrogen reductions may subsequently offer the option of reducing sulfur or nitrogen
5 only nitrogen reduction will offer protection
6 either sulfur or nitrogen reductions may offer protection
7 ecosystem protection where critical loads are not exceeded

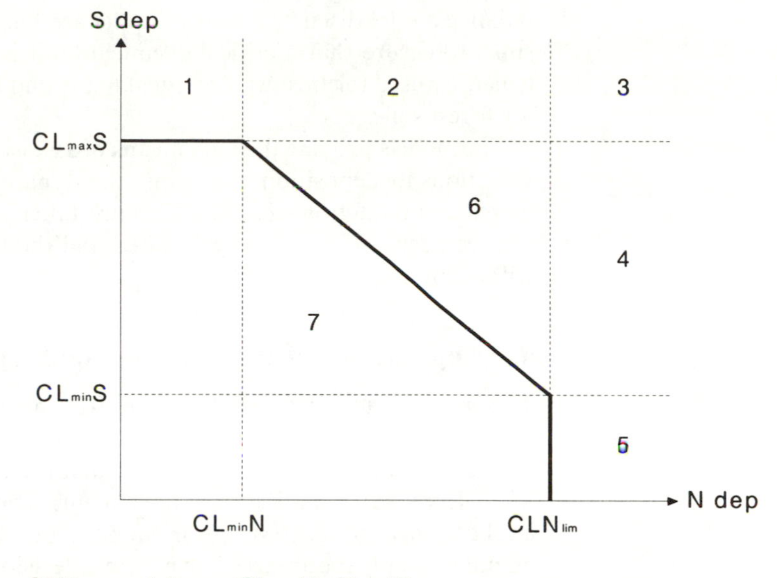


S dep = sulfur deposition
N dep = nitrogen deposition
$CL_{max}S$ = maximum critical load of sulfur
$CL_{min}S$ = minimum critical load of sulfur
$CL_{min}N$ = minimum critical load of nitrogen
$CL_{max}N$ = maximum critical load of nitrogen
$CL_{nut}N$ = critical load of nutrient nitrogen
$CLN_{lim}$ = limiting critical load of nitrogen

$CLN_{lim} = \text{minimum}(CL_{max}N, CL_{nut}N)$

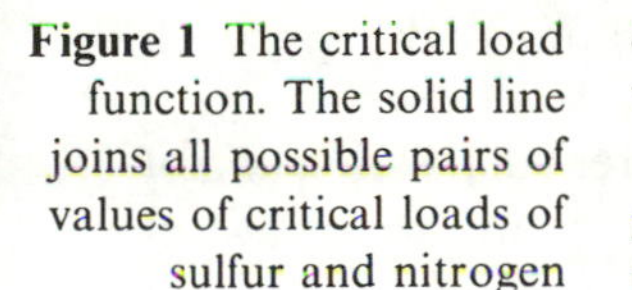

**Figure 1** The critical load function. The solid line joins all possible pairs of values of critical loads of sulfur and nitrogen

and hence the response of the system to deposited acidity; these include mineral weathering, ion exchange and uptake of base cations by forest growth.

There are three main families of models which have been used. (i) The MAGIC model[24] and its developments MAGIC-WAND[25] and MERLIN;[25] the original MAGIC model was designed to examine acidification of waters at the catchment scale and has been applied to individual catchments and at the regional scale. The MAGIC-WAND and MERLIN developments incorporate more sophisticated handling of N dynamics. (ii) The SAFE model[25] is a dynamic version of the PROFILE model,[21] which relates to equilibrium conditions, developed at the University of Lund. It can be applied at the soil profile or catchment scale and differs from the other two families of models in that weathering rate is calculated from input data on soil mineralogy, surface area and moisture contents. (iii) The SMART model[26] and the related RESAM model[27] are mainly designed to examine regional trends in acidification and the RAINS model has been extensively used in mapping of critical loads at the European scale. The related NUCSAM model,[28] which considers nutrient cycling in more detail, has been linked to FORGRO,[29] a forest growth model to give a linked soil–tree model. The models are used to calculate the critical load in a similar way to the mass

[24] B. J. Cosby, G. M. Hornberger, R. F. Wright and J. N. Galloway, *Water Resour. Res.*, 1985, **21**, 51.
[25] R.C. Ferrier, W. de Vries and P. Warfvinge, in *Mapping and Modelling of Critical Loads for Nitrogen—a Workshop Report*, ed. M. Hornung, M.A. Sutton and R.B. Wilson, Institute of Terrestrial Ecology, Edinburgh, 1995, p. 97.
[26] W. de Vries, M. Posch and J. Kamari, *Water, Air, Soil Pollut.*, 1989, **48**, 349.
[27] W. de Vries, J. Kros and C. van der Salm, *Water, Air, Soil Pollut.*, 1994, **75**, 1.
[28] J. E. Groenberg, J. Kros, C. van der Salm and W. de Vries, *Ecol. Modell.*, 1995, **83**, 97.
[29] G. M. J. Mohren, I. I. M. Jorritsma, J. P. G. M. Florax, H. H. Barilink and J. R. van der Veen, *FORGRO 3.0. A Basic Forest Growth Model: Model Documentation and Listing*, Report 524, Research Institute for Forest and Urban Ecology, Wageningen, 1991.

balance model discussed above. Inputs are balanced with internal sources and sinks to ensure that a critical chemical limit is not exceeded; an ANC limit is generally used for freshwater critical loads and the base cation:aluminium ratio for forest soils.

The models provide the only means of assessing the response of ecosystems to variations in deposition over time, the dynamics of the response to emission increase or reduction scenarios. The time taken to reach a new equilibrium under changed deposition can be calculated and the time at which predicted changes will occur.

## 4 Applications of the Critical Load Approach

The critical loads approach has been used most extensively in Europe, where it has been incorporated into the development of policy and the setting of emission targets by the UNECE and a number of national governments, including the UK. Most European countries now have teams working on critical loads and their application, in order to supply input data for European scale maps, even if they do not yet apply the approach in national decision making. The application of the approach is also being explored in a number of Asian countries, including China and Japan; collaboration between Asian and European research groups in this area of work is being supported by the World Bank. Application of the approach in South Africa has recently been reported.[30]

The approach has been used to assess the impacts of current deposition and of future deposition scenarios. The concept was first developed in the context of transboundary air pollution but its use at smaller scales, *e.g.* to individual catchments, is also being explored, although questions have been raised about the limitations of the approach at these scales.[31] The following section illustrates some of the applications at different scales and with different deposition scenarios.

### *Combined Receptor and Percentile Critical Loads*

The initial emphasis of national mapping programmes was on the production of single receptor maps, for example soils, freshwaters, selected vegetation habitats; maps of this type have been discussed in outline above. Although these single receptor maps are still important, there has been a move towards combined receptor maps. There has also been a parallel move to the production of percentile maps, with a 5-percentile approach being used most commonly.[32] A percentile critical load is the value that will protect a given percentage of the ecosystems in a grid square. Thus, a 5-percentile critical load is the value of the critical load that will protect 95% of the total area of ecosystems in a grid cell. The

[30] A. M. van Tienhoven, K. A. Olbrich, R. Skoroszewski, J. Taljarrdand and M. Zunckel, *Water, Air, Soil Pollut.*, 1995, **85**, 2577.

[31] M. Hornung, B. Reynolds, M. Brown and D. C. Howard, in *Exceedances of Critical Loads and Levels: Spatial and Temporal Interpretations of Elements in Landscapes Sensitive to Atmospheric Pollutants*, ed. J. Schneider and G. Soya, Federal Environment Agency, Vienna, 1996, p. 269.

[32] J.-P. Hettelinge, W. de Vries, W. Schopp, R. J. Downing and S. A. M. de Smet, in *Calculating and Mapping of Critical Loads in Europe*, ed. R. J. Downing, J.-P. Hettelingh and P. A. M. de Smet, RIVM, Bilthoven, 1993, p. 6.

**Table 2** Ecosystem type, ecosystem critical load and area for example calculation of 5-percentile critical load

| *Ecosystem* | *Critical load* | *Ecosystem area*/km$^2$ *and % of total ecosystem area* |
|---|---|---|
| Acid grassland | 0.5 | 10 (4.2%) |
| Heathland | 0.6 | 80 (33.3%) |
| Woodland | 0.8 | 150 (62.5%) |

approach requires, for each grid cell, the critical loads for each receptor being considered and the area of the cell occupied by each receptor. A cumulative distribution is then produced for the critical loads for each square and to derive the value that will provide the required level of protection. For example, using the data for the three ecosystems in Table 2, the receptor with the lowest critical load, the acid grassland, occupies only 4.2% of the total ecosystem area in the grid cell. Setting the critical load at the value of 0.5 would provide protection for more than 95% of the total area of the selected ecosystems. The area of the ecosystem(s) with the next lowest critical load is therefore added, the heathland, and its critical load applied. The new total is above 5% of the total and so the critical load would be set to 0.6.

The current UK percentile maps are being produced by combining critical loads for individual receptors calculated using a variety of approaches. Thus, the simple mass balance model is used to calculate critical loads for woodland, the empirical critical loads for soils are used to derive the critical load of acidity and S for other habitats, the diatom critical loads are used for critical loads of acidity and S for surface waters, the N mass balance model is used for critical loads of N for woodland and empirical critical loads of N for other habitats.

## *Mapping of Critical Loads at the European Scale*

Under the UNECE Convention on Long Range Transboundary Air Pollution, production of European maps is carried out from a co-ordination centre (the CCE) in Bilthoven, the Netherlands. The maps are currently based on the EMEP (European Monitoring and Evaluation Programme) grid of 150 km cells for which deposition data are modelled at the EMEP centre in west Norway,[33] using a long-range transport model. The mapping will move to 50 km grid cells over the next year and as deposition data become available at this scale. National critical load data are submitted to the co-ordination centre where they are merged; data gaps, for countries which do not submit their own national data, are filled by a modelling exercise in the co-ordination centre using the RAINS model. The UK submits data for each 1 km grid square of the National Grid but other countries provide data at different resolutions; the use of the EMEP grid cells allows for European maps to be generated at a single resolution which also matches the deposition data available at the European scale.

The co-ordination centre formerly produced maps of the critical deposition of S, derived from the critical load of acidity using the 'sulfur fraction', and of N. The acidity critical load was divided into sulfur and nitrogen fractions on the basis of

[33] J.-P. Tuovinen, K. Barrett and H. Styve, *Transboundary Acidifying Pollution in Europe*, Meteorological Synthesising Centre-West, Oslo, 1994.

the current deposition loads of the two pollutants.[34] The maps of the critical deposition of S were used in the negotiation of the Second Sulfur Protocol in combination with the sulfur deposition data produced by EMEP. The deposition data could be scaled with respect to emissions from individual countries across Europe, thus allowing deposition scenarios to be defined for any combination of national emissions and then compared with the critical loads. The discussions in the UNECE also considered the costs of abatement to achieve given emission and deposition scenarios, using data on the costs of abatement for individual sources. The linking of modelled deposition, critical loads and abatement costs has become known as integrated assessment modelling.

The negotiations actually used a 5-percentile critical load map for sulfur but the integrated assessment modelling indicated that these critical loads could not be achieved within realistic abatement costs. A gap closure approach was eventually used to set the targets for emission reductions. Thus, target loads were set which reduced the gap between 1980 emissions and the 5-percentile critical loads in any EMEP grid cell by 60% by the year 2010. This clearly did not provide protection for all ecosystems in all areas but greatly reduced the area over which the critical load was exceeded. Thus, using 1990 deposition, 24 EMEP grid squares in Europe, mainly in a belt from the UK to Poland and the Czech and Slovak Republics, were mapped as having large exceedances ($<1.5$ keq ha$^{-1}$ yr$^{-1}$). Modelled deposition data incorporating the implementation of the protocol predicts that only two squares will have large exceedances in 2010.[1]

The mapping approach used by the CCE has now changed, reflecting the move to a critical load for S plus N and the use of the critical loads function. Maps are now produced for $CL_{max}S$, $CL_{min}N$, $CL_{max}N$ and $CL_{nut}N$. Countries submit data for each ecosystem covering more than 5%, and for which critical loads have been calculated, in each of their national mapping cells. In practice, most countries only submit data for one ecosystem, generally forests, although the Scandinavian countries also submit data for freshwaters. At the CCE, the data are used to create a frequency distribution for each EMEP grid square of critical load values and associated ecosystem areas. Percentile critical load values are computed for each EMEP grid square using the approach outlined above for the UK national maps. The critical load function approach is used at the CCE to generate protection isolines and to map areas where (i) N or S reductions are required, (ii) mandatory S reductions are required, (iii) mandatory N reductions are required or (iv) mandatory S and N reductions are both required.

## *Exceedance and Calculation of Emission Reduction Requirements*

Comparison of the critical load map/database for a given receptor, or a combination of receptors, with measured or modelled annual deposition data for the pollutant of interest will show if, and by how much, the critical load is exceeded. The amount of exceedance is equivalent to the reduction in deposition required to provide protection for the chosen receptor. Production of national or regional exceedance maps requires a map/database of the stock at risk/distribution

[34] J. Hall, M. Brown, H. Dyke, J. Ullyett and M. Hornung, *Water, Air, Soil Pollut.*, 1995, **85**, 2521.

of the receptor of interest, a critical load map/database for the stock at risk and a map/database of deposition for the pollutant of interest. Overlaying the critical load map with the deposition map shows the spatial patterns of exceedance and hence where deposition reductions need to be targeted.

*Exceedance by Current Deposition.* Figures 2(a) to 2(c) show (a) the empirical critical load map of acidity for soils of the UK, (b) the modelled annual deposition of sulfur based on data for the period 1992 to 1994 and (c) the exceedance map produced by overlaying the other two maps. In this instance the dominant soils in each 1 km grid cell of the National Grid are considered the stock at risk. The deposition data are derived using measured data from a national network of precipitation concentration and gas monitoring stations.[35,36] The critical load map is based on a 1 km grid and the deposition data on a 20 km grid; the deposition for a given 20 km grid cell is applied to each 1 km cell of the soil map within the 20 km square. The exceedance map indicates that the critical load is exceeded in 88 077 km squares, or 37% of the UK, with the main components of the areas of exceedance being in the uplands of the north and west. A reduction in deposition of $>1.0\,keq\,ha^{-1}\,yr^{-1}$ would be required in some 3% of the UK to provide complete protection.

*Evaluation of Exceedance by Alternative Deposition Scenarios.* The calculation of deposition fields based on measurements is constrained by the availability of the measurement data. In the UK the number of monitoring sites and the protocols for running these have varied through time, with a national precipitation composition network only being established in 1986.[37] Assuming that emissions data are available, however, the use of modelled deposition fields extends the time range over which exceedances can be estimated, both in the past and into the future to explore the possible effects of emissions reduction policies. Here we describe the application of the HARM model to two periods for which we have no measurements: 1970 and 2010.

HARM (Hull Acid Rain Model)[38] is a receptor orientated Lagrangian statistical model which estimates annual concentrations and depositions of a range of pollutants across the UK. The model tracks the changing composition of air parcels moving towards designated receptor sites. Emissions of $SO_2$, $NO_x$ and $NH_3$ are held in 150 km × 150 km grids across the EMEP area and, with the addition of HCl, in 20 km × 20 km grids across the UK. These emissions are disaggregated by source type. UK $SO_2$ emissions, for example, are held for power stations, oil refineries, other industrial sources and low levels sources. The model assumes instantaneous mixing throughout the air parcel and source height is not taken into consideration. The model represents the coupled chemistry of these

[35] RGAR, *Acid Deposition in the United Kingdom 1986–1995. 4th Report of the United Kingdom Review Group on Acid Rain*, Department of the Environment, Harwell, 1997.

[36] R. I. Smith, D. Fowler and K. Bull, in *Acid Rain Research: Do We Have Enough Answers?*, ed. G. J. Heij and J. W. Erisman, Elsevier, Amsterdam, 1995, p. 175.

[37] RGAR, *Acid deposition in the United Kingdom 1986–88. 3rd Report of the United Kingdom Review Group on Acid Rain*, Department of the Environment, Warren Spring, 1990.

[38] S. E. Metcalfe, D. H. Atkins and R. G. Derwent, *Atmos. Environ.*, 1989, **23**, 2033.

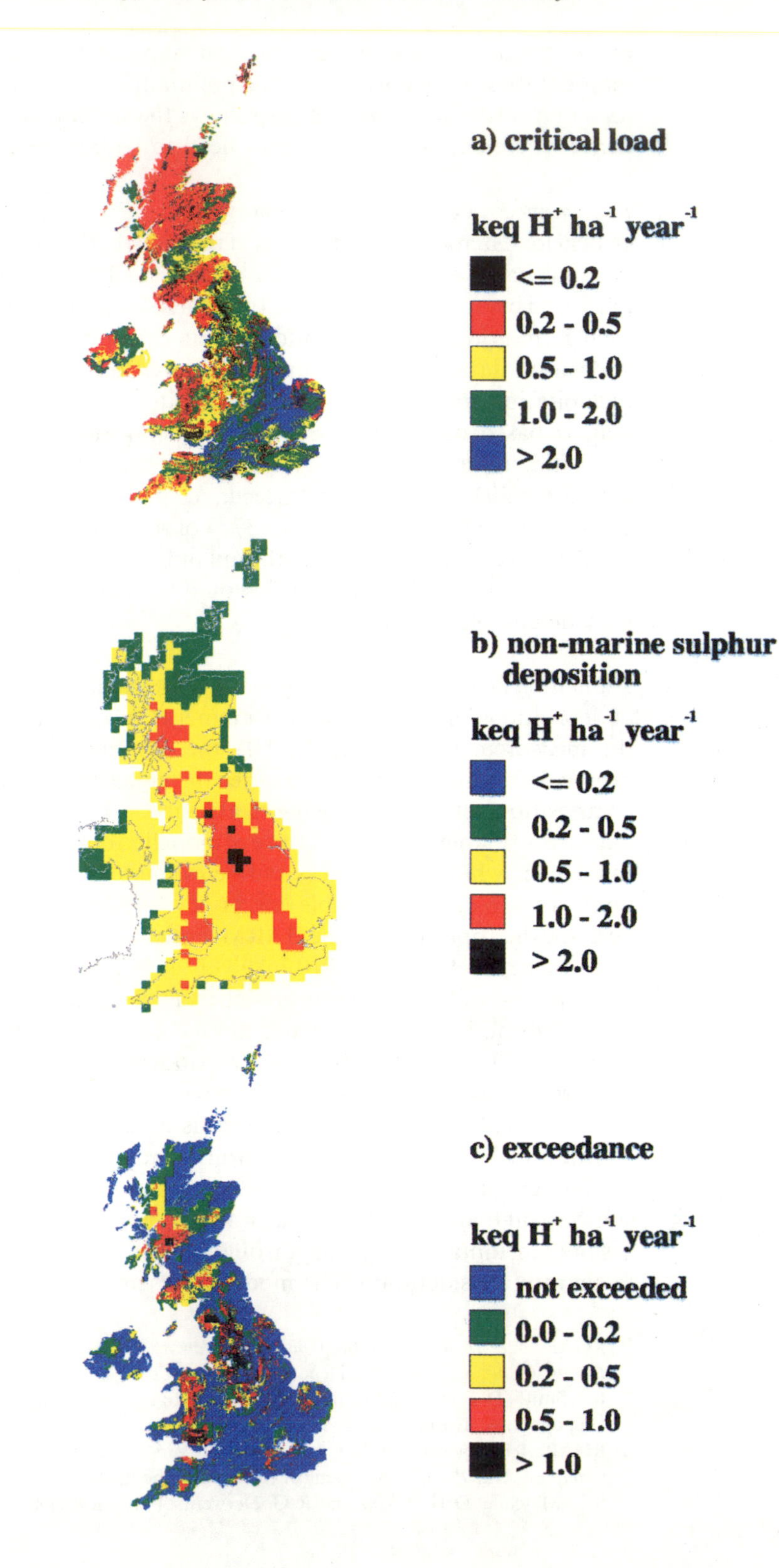


**Figure 2** Exceedance of critical loads of acidity for soils in the UK by current deposition. (a) Empirical critical loads of acidity for soils, (b) mean annual deposited sulfur 1992–1994, (c) exceedance of the critical loads for soils—derived by combining the data represented by (a) and (b)

pollutants, assuming a constant supply of OH radicals and setting an initial $O_3$ concentration of 30 ppb. The timestep is 2 minutes and the air parcels have a maximum travel time of 96 hours. A highly simplified representation of meteorological conditions is employed assuming constant wind speed, straight line trajectories, a constant boundary layer height (800 m) and constant drizzle. Modelled wet deposition includes a simplified representation of the seeder–feeder process which makes a significant contribution to deposition in upland areas.[39] Details of the sulfur part of the model and a comparison with data from the UK's monitoring networks are given in Metcalfe *et al.*[40]

HARM is used extensively for source attribution work and to assess the impacts of changes in the magnitude and spatial distribution of emissions. UK emissions of $SO_2$ peaked in 1970 at about 5.8 million tonnes,[41] declining to just over 2 million tonnes by 1994. The current reduction plans set a target of 980 000 tonnes for 2010. As no national maps of acidic deposition (from measurements) are available before 1986, the only way to explore the possible impacts of these changes in emissions on depositions and exceedances is to use a model such as HARM. Although deposited acidity is a combination of a number of pollutants, it is simplest to assume that all acidity is S and to compare modelled S deposition with critical load for acidity. The exceedance maps which result from such an exercise must be regarded as a minimum estimate of ecosystem damage from acid deposition. Although a total acidity approach can be adopted and modelled S and N deposition used with the critical loads function (see below), this requires a number of additional assumptions to be made which are beyond the scope of the present discussion.

Figure 3 shows HARM modelled S deposition for 1970 and 2010 and the resulting exceedance maps of the 1 km empirical critical load for the acidity of soils. The 1970 UK disaggregated inventory is taken from Smith and Jeffrey[41] and emissions for the remainder of the EMEP area are those used by Mylona.[42] UK $SO_2$ emissions for 2010 are based on information provided by the Environment Agency and its equivalent organizations in other parts of the UK for power stations and oil refineries and on DTI estimates for other sources. The total emission is 905 000 tonnes, less than the published current reduction plan to meet the terms of the 1994 Oslo protocol. Emissions estimates for the EMEP area were supplied by Imperial College, London. As might be expected, the changes in S deposition and critical load exceedance are quite dramatic. Total, area weighted, modelled S deposition to the UK in 1970 is 579.5 ktonnes, while in 2010 this figure has fallen to 138 ktonnes S. For comparison, the estimate for 1992–94 based on measurement data is 366 ktonnes S.[35] Figure 3(a) shows depositions in excess of 1.5 keq (24 kg S $ha^{-1}$) over much of England and Wales and in parts of western and southern Scotland. The model indicates that dry deposition predominated over wet deposition (341.5 ktonnes dry, 238 ktonnes wet), probably reflecting the importance of UK sources. In 2010, however (Figure 3b),

[39] A. J. Dore, T. W. Choularton and D. Fowler, *Atmos. Environ.*, 1992, **26A**, 1375.

[40] S. E. Metcalfe, J. D. Whyatt and R. G. Derwent, *Q. J. R. Meterol. Soc.*, 1995, **121**, 1387.

[41] F. B. Smith and G. H. Jeffrey, *Atmos. Environ.*, 1975, **9**, 643.

[42] S. Mylona, *Trends of Sulphur Dioxide Emissions, Air Concentrations and Depositions of Sulphur in Europe since 1880*, EMEP/MSC-W Report 2/93, Norwegian Meteorological Institute, Oslo, 1993.

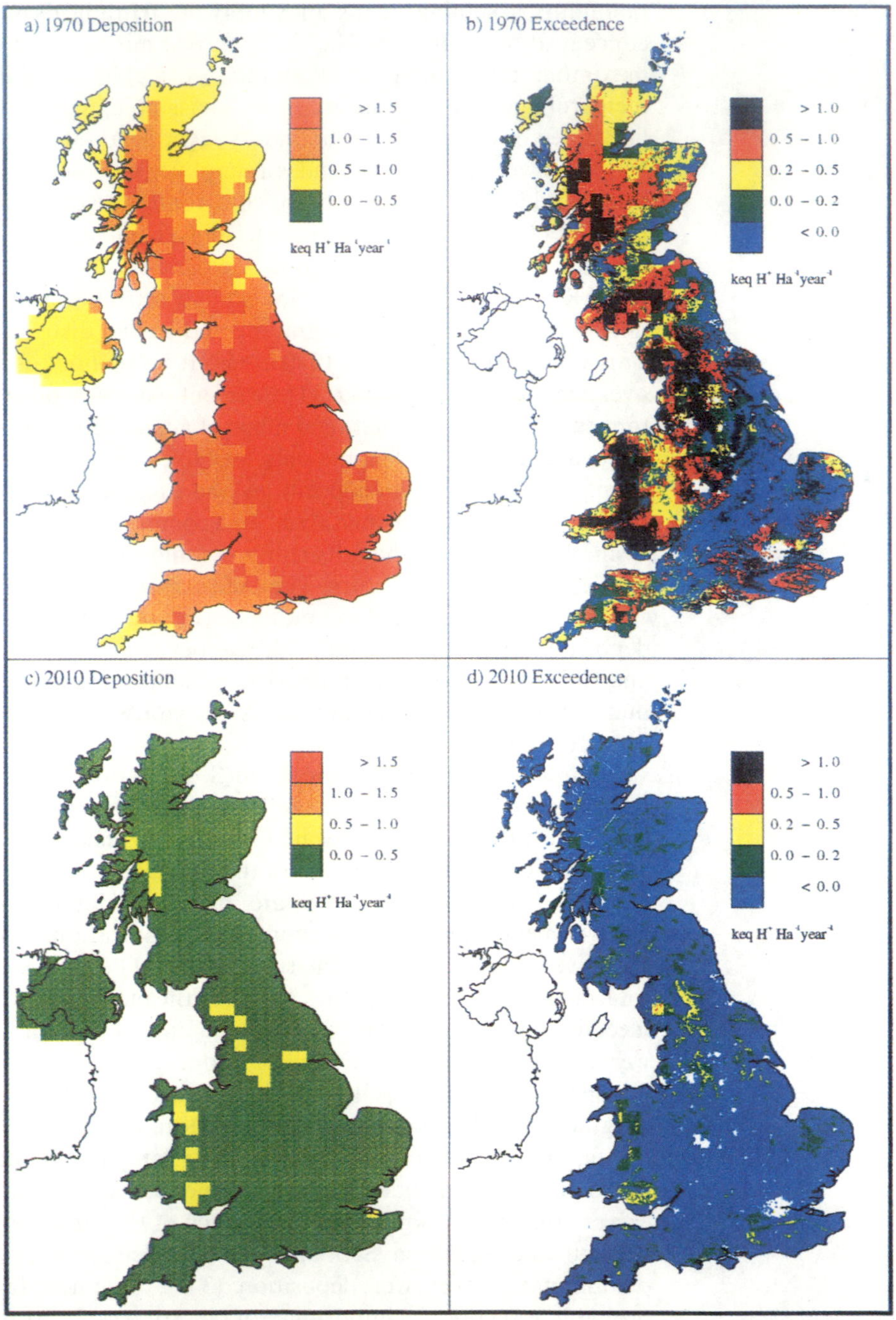

**Figure 3** Exceedance of critical loads of acdity for soils in the UK by 1970 and 2010 sulfur deposition. (a) 1970 deposition, (b) 1970 exceedance, (c) 2010 deposition, (d) 2010 exceedance

the maximum modelled deposition to any 20 km grid square is 0.59 keq (9.4 kg S $ha^{-1}$). The balance of wet and dry deposition has also changed, with wet exceeding dry. It is believed that this reflects the greater importance of long-range transport of S. Recent budget estimates by the Review Group on Acid Rain, based on data from the monitoring networks, suggest that in 1987 dry deposited S was greater than wet deposited, whereas by 1992–94 wet deposition was greater than dry. Dry deposition appears to be declining more rapidly than wet, in response to emissions reductions as suggested by the HARM output.

As described above, exceedance maps (Figures 3b and 3d) have been generated by overlaying the 20 km S deposition fields on to the 1 km critical load values. Inevitably, this masks what could be very considerable variability in deposition at the sub-20 km scale, particularly in areas of complex topography. At present, even deposition figures based on measurements are only derived at the 20 km × 20 km scale since it is believed that some processes (*e.g.* seeder–feeder enhancement) cannot yet be applied at a finer scale. Statistical estimates of the misrepresentation of critical load exceedance that might result from the mismatch in spatial scales between deposition and critical load have been discussed.[36] The effects of the 20 km deposition can be seen on the exceedance maps in Figure 3. Using the 1970 modelled deposition, critical loads were exceeded in some 64% of the UK, with about 18% in the highest exceedance category. Areas of high exceedance cover parts of south-east England, most of Wales and the uplands of northern England and Scotland outside the central valley. The pattern of exceedance in 2010 is quite different. Critical load exceedance is confined to about 8% of the UK, with no 1 km squares in the highest exceedance category and only 0.01% in the 0.5–1.0 exceedance category. Areas of exceedance persist across the country, but with the highest levels in Cumbria, the Pennines and south Wales. This approach suggests parts of the country where proposed emissions reductions will bring depositions below critical loads and hence places where signs of system recovery might be expected. Equally, it is clear that, in many highly acid sensitive upland areas, exceedance is likely to persist and, with it, ecosystem damage. Other work with HARM has suggested that the protection of many areas of western and northern Britain would require reductions in $SO_2$ emissions from other EMEP countries beyond those agreed as part of the Oslo protocol.

It is clear that the combination of models of the transport and deposition of air pollutants and critical loads maps offers a useful tool in policy formulation. Indeed, this has been the approach adopted within the UNECE since 1988.[43] At the European scale the EMEP model provides estimates of depositions of S and N at a scale of 150 km × 150 km[44] which can then be compared with critical load data held at the Coordinating Center for Effects in the Netherlands. This scale of approach is rather coarse for the UK; hence the use of a deposition model specifically designed for the UK and of the national critical load data set. Combinations of the deposition fields from different models and critical load data sets compiled in slightly different ways will yield varying patterns of exceedance. The methodology, however, provides a more satisfactory and politically acceptable way of determining emissions reductions than the blanket approach adopted in early UNECE protocols. At the local scale in the UK, it was also part of the approach adopted by the Environment Agency in the latest round of

[43] R. G. Derwent and R. B. Wilson, in *Critical Loads: Concept and Applications*, ed. M. Hornung and R. A. Skeffington, HMSO, London, 1993, p. 5.

[44] *Transboundary Air Pollution in Europe*, ed. K. Barrett and E. Berge, EMEP/MSC-W Report 1/96, Norwegian Meteorological Institute, Oslo, 1996.

allocations of emissions from power stations in England and Wales. Although regarded by some as a rather blunt instrument, using reductions in critical load exceedance as a policy target appears to be an established part of environmental policy across Europe.

## *Consideration of Emissions from Individual Sources*

Critical load maps/databases for specific receptor types can be combined with modelled deposition data based on emission data for individual sources to assess the contribution of the individual sources to exceedance of the critical load for the chosen receptor. Thus, the contributions of individual UK sources, and of deposition derived from mainland European sources, have been calculated for sulfur deposition at Sites of Special Scientific Interest (SSSIs) in England and Wales where the critical load was exceeded.[45] They used modelled sulfur deposition data for 1993 and 2001, derived from the HARM model,[38] in combination with the empirical critical loads map of acidity for soils and a boundary data set for SSSIs; the data were manipulated within a GIS. It was calculated that 'European emissions' contribute 39% of the deposition to the exceeded SSSI areas using 1993 data and 43% of the deposition to these areas using 2001 deposition. The largest of the UK sources, the Drax power station, is calculated to contribute 3.8% of the deposition to the exceeded areas in 1993 and 1.8% in 2001, reflecting the operation of emission control equipment.

## *Consideration of Interactions between Emission Reductions for S and N: Application of the Critical Load Function*

As noted earlier, application of the critical load function allows the simultaneous consideration of the requirements for reductions in emissions of S and N. It can be used to determine whether reductions in both S and N deposition are essential to provide protection or whether there is the option of reducing one or the other (Figure 1). When combined with modelled deposition data, the approach has considerable potential for scenario analysis of deposition reduction options required to achieve critical loads. Thus, for example, modelled S and N deposition data, derived using the HARM model,[38] have been used in combination with critical load maps for shrub heath in the UK to assess exceedance and emission control options to provide protection for this habitat.[40] The HARM model calculates deposition from emission data for major UK sources. With current deposition, the critical load of S + N is not exceeded in 399 out of 720 20 km grid cells. The majority of the remaining 321 grid squares are predicted to require reductions in S deposition, with the majority requiring primary decreases in S before there is an option of reducing S or N. A minority of grid squares require a primary reduction in N deposition. However, following

[45] M. Brown, H. Dyke, S. M. Wright, R. A. Wadsworth, K. R. Bull, A. Farmer, S. Barham, S. E. Metcalfe, D. Whyatt and C. Powlesland, *Water, Air, Soil Pollut.*, 1995, **85**, 2589.

implementation of the Second Sulfur Protocol, it is predicted that the number of protected squares will rise to 507 and that the majority of exceeded squares will then require primary reductions in N deposition.

The critical load function approach, as used above, can be extended by considering the limiting values of S and N deposition needed to achieve protection. For any point in the area of exceedance, a maximum S reduction can be defined which will help achieve the critical load. This value will be associated with a residual N reduction. The alternative approach is to maximize N reductions. Using modelled current deposition for the UK it can be shown that there is scope for further reductions of sulfur, thus reducing the reduction in N required to provide protection. However, reducing N deposition from current levels substantially reduces any need for further S reductions. The modelled 2010 deposition, incorporating the Second Sulfur Protocol, shows that the scope for further reductions in S are much more limited. It may prove easier to consider N reductions which, if maximized, would result in only minimal S reductions being required to meet critical loads.

## *Application of Dynamic Models*

Dynamic models have been used in two main ways in the context of critical loads: (i) to examine the time to exceedance, or of recovery to non-exceedance, mainly by modelling changes in the parameters used to set critical chemical limits, for example ANC zero for freshwaters or the relevant base cation: aluminium ratio for vegetation–soil systems; and (ii) to evaluate deposition–land use interactions and the implications for critical load exceedance. The models are usually applied to specific sites but regionalized versions have also been developed which have been applied to assess, for example, the proportion of lakes in a region in which the critical ANC value, and hence the critical load, will be exceeded under a given deposition scenario.[46]

The pollutant–land use interactions have mainly considered the impact of afforestation and of replanting forests following clear felling. The replacement of upland grassland or shrub heath with coniferous plantations can result in an increase in pollutant deposition since forest growth and harvesting involves a removal of base cations, and hence a potential reduction in the buffering capacity of the soil system. The increased deposition and uptake can be incorporated into the models which can then be used to assess whether afforestation of part or all of a catchment in sensitive areas of the uplands can result in a shift from non-exceedance of the critical load for the surface waters to exceedance. The models have also been used to assess the impact of replanting after harvesting as opposed to allowing the site to revert to the original grassland or moorland.[46] In this situation the models can be used in a management context and to support decision making, with the critical load as the environmental quality threshold.

[46] A. Jenkins, in *Critical Loads of Acid Deposition for United Kingdom Freshwaters*, Institute of Terrestrial Ecology, Edinburgh, 1995, p. 35.

## 5 Summary and Conclusions

The critical load concept provides a valuable addition to the range of tools available for setting emission targets for pollutants. Its main strengths are that it is receptor based and able to allow for the spatial variation in the sensitivity of receptors to a given pollutant and the spatial relationship between sources and receptors. The approach has been most widely applied in the context of ecological receptors, *e.g.* forests, soils, surface waters and natural vegetation communities, but has also been explored for building materials. For the ecological receptors, a number of alternative methodologies have been developed for the calculation and mapping of critical loads for sulfur and nitrogen. These methodologies, ranging from empirically assigned values to dynamic models, vary in complexity and the choice of method is determined by data availability and the aim of the particular exercise. Methodologies are also currently being explored for heavy metals and persistent organic pollutants.

The critical loads approach has been incorporated into policy development, and target setting at national and European scales. More than 10 European countries have now produced national critical load maps for acidity and sulfur and the approach is being tested in North America, South Africa, China, Japan and a number of other Asian countries. The negotiation of the Second Sulfur Protocol, under the auspices of the UNECE Convention on Long Range Transboundary Air Pollution, incorporated the use of critical loads at the European scale and they are also being used in discussions on a nitrogen protocol.

However, it would be wrong to convey the impression that the current methodologies represent definitive versions. The available methodologies continue to be refined using the results from relevant research and trial applications. Further research is needed on the underlying assumptions and principles. Thus, for example, the concept assumes a threshold load of the pollutant below which the receptor will not be damaged. Biological systems rarely respond in this way to stress. In the application of mass balance based approaches, the choice of biological indicator and the setting of the associated critical chemical limit are important. Appropriate indicators need to be identified for a wider range of habitats and critical chemical values set for those indicators. The methodologies also require testing in a wider range of environmental conditions. The consequences of the exceedance, and magnitude of exceedance, of the critical load for a given receptor needs to be assessed. Approaches need to be developed for expressing the uncertainties in calculated critical loads, particularly in mapped outputs.

## 6 Acknowledgements

The authors' own work on critical loads has been supported by the UK Department of the Environment and the Natural Environment Research Council. The authors are grateful to their colleagues on the UK Critical Loads Advisory Group and its various sub-groups, on whose work they have drawn, and to Dr Duncan Whyatt for producing Figure 3.

# California's Approach to Air Quality Management

ALAN C. LLOYD

## 1 Introduction

Air pollution control in California dates from the mid to late 1940s, when the Air Pollution Control District in Los Angeles County was created. Then, in 1967, the effort shifted to the state level with the creation of the California Air Resources Board (CARB) and air pollution control districts throughout the state. These districts were created along geographic and topographic lines as well as on county bases and their work complements that of the CARB. Currently there are 15 air basins and 34 air pollution control districts and air quality management districts in California, the largest of which is the South Coast Air Quality Management District (SCAQMD).[1] The SCAQMD covers approximately 12 000 square miles and encompasses all of Los Angeles, Orange and Riverside Counties and the non-desert portion of San Bernardino County. The population is approximately 14 million, which is just under half of the state population of over 32 million.

Under the federal Clean Air Act of 1970, California was granted special status because of the severity of its problems, particularly in the Los Angeles area. The state is allowed to set its own, more stringent, emissions standards for vehicles, a policy which continues today.

While the CARB is primarily responsible for monitoring and regulating emissions from mobile sources, since they occur throughout the state, the local air districts are responsible for defining and implementing air pollution control measures for stationary and area sources. Each control or management district is able to set stationary source emission standards that may differ from the state or federal standards. This is typically the case in the larger air quality management districts, such as the SCAQMD and the San Francisco Bay Area Air Quality Management District. The US Environmental Protection Agency (US EPA) is responsible for standards relating to sources which cannot be controlled by the state, such as off-road sources and emissions from trains and aircraft.

California has seen a significant improvement in its air quality in the past

[1] CARB, *California's Air Quality Data for 1995*, California Air Resources Board, Sacramento, 1995.

decade.[2] This improvement is a result of an extensive, tough control program based generally on sound science.

This article reviews the trends in emissions and subsequent air quality resulting from implementations of control strategies in California during approximately the last 20 years. A discussion of the scientific and technical rationale for the approaches taken to control air pollution includes the mechanisms utilized to implement the various control measures. The more recent efforts to encourage cleaner burning fuels and to require zero-emission light- and medium-duty vehicles are discussed. The latter part of the article addresses some of the shortcomings of the management program; the interaction between economic growth and air pollution controls is briefly discussed because of their significant interplay during the early 1990s, when California experienced its first extended economic recession.

## 2 Emissions and Air Quality Trends

Air quality in California, tracked at approximately 160 stations statewide, has improved steadily and significantly over the last 20 years. For example, Figures 1 and 2 show the statewide trend in air quality for ozone and benzene, respectively, during this period. Similar trends are observed for carbon monoxide and nitrogen dioxide. These data indicate that air quality has improved dramatically despite substantial increases in population and vehicle miles traveled. The Los Angeles area, included in the SCAQMD, shows the most dramatic improvement in the state. This is illustrated in Figure 3, which shows the percent change for the various parameters between 1982 and 1994.[2]

Table 1 shows a comparison of the annual basin days that various ozone standards were exceeded for the period 1976 to 1996.[3] No matter which statistic is examined, one can see a dramatic improvement in ozone concentrations in the SCAQMD during this period. For example, the number of days that ozone levels

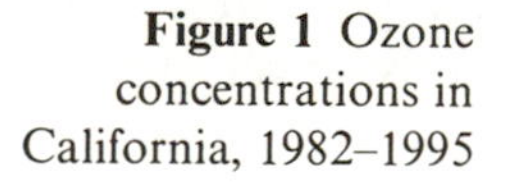

**Figure 1** Ozone concentrations in California, 1982–1995

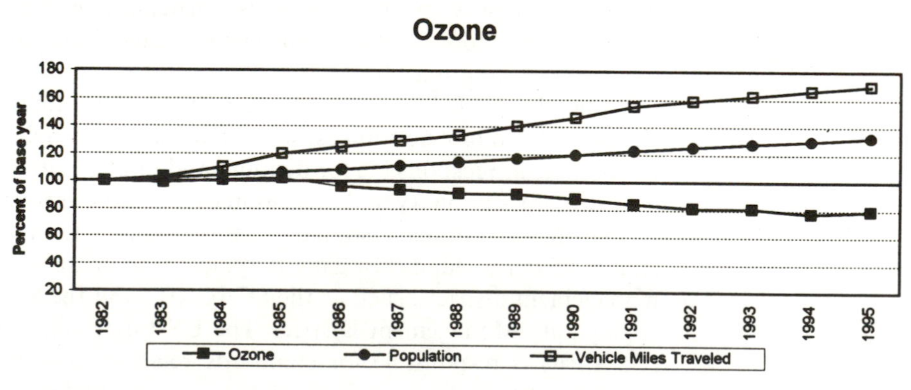


[2] CARB, *California's Air Quality... The Success Story*, California Air Resources Board, Sacramento, 1996.

[3] SCAQMD, *Preliminary Analysis of* 1996 *Air Quality*, South Coast Air Quality Management District, Diamond Bar, CA, 1997.

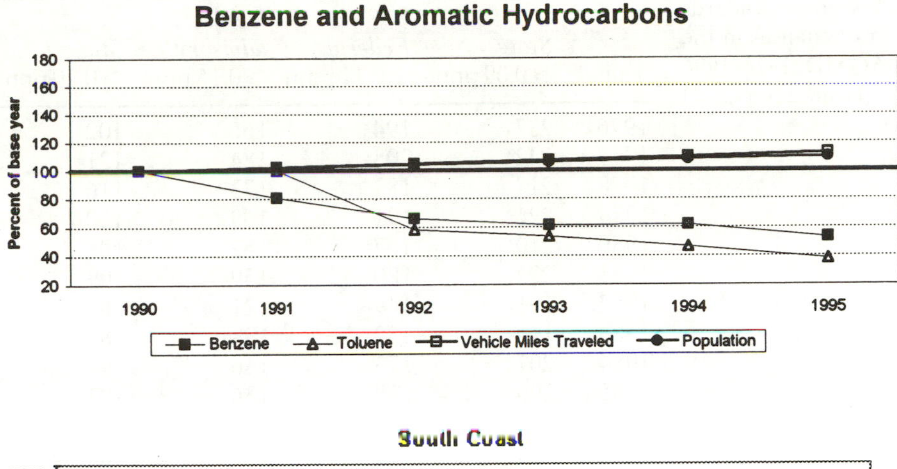


**Figure 2** Aromatic hydrocarbon concentrations in California, 1990–1995

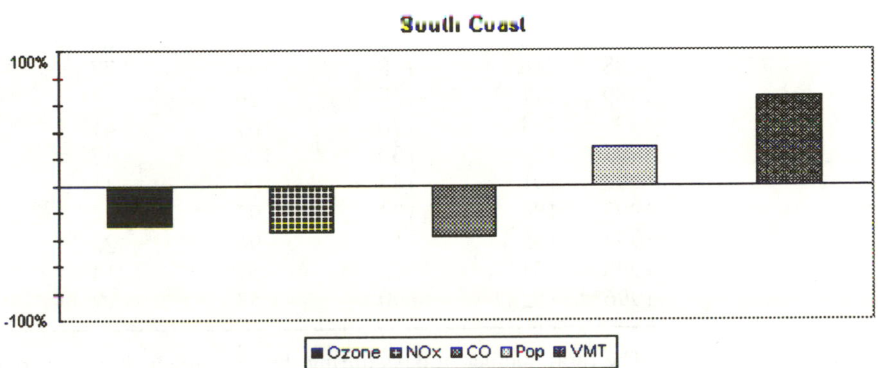


**Figure 3** Percent changes in air quality parameters, 1982–1994

were greater than 0.20 ppm decreased from 102 to 7 during this time. The number of days in which the health advisory standard of 0.15 ppm was exceeded decreased from 166 to 53. It is clear that control measures have been working very well in spite of continued population growth and an increase in vehicle miles traveled.

However, one must also recognize that the federal and state ozone standards of 0.12 ppm and 0.09 ppm, respectively, are still exceeded on 90 and 150 days, respectively. While substantial progress has been made, further controls and reductions are necessary to attain these health-based air quality standards. In addition, the expected new eight-hour standard for ozone will be more difficult to attain than the current ozone standards illustrated in Figure 4. There is an economic as well as legal impetus to attain these standards; Hall *et al.*[4] estimated that the cost of air pollution, in terms of health effects alone, was at least US $9 billion per year in the SCAQMD. As mentioned below in Section 9, additional challenges will be posed if the new eight-hour ozone and fine particle standards are promulgated.

On-road motor vehicles contribute the largest fraction of emissions in California. They comprise approximately 42% of the reactive organic gases

[4] J. Hall, A. M. Winer, M. T. Kleinman, F. W. Lurmann, V. Brajer and S. D. Colome, *Science*, 1992, **255**, 812.

**Table 1** Ozone standards exceedances in the SCAQMD, 1976–1996 (basin days/year*)

| *Year* | *State* >0.09 ppm | *Federal* ⩾0.12 ppm | *Health advisory* ⩾0.15 ppm | *Stage I* ⩾0.20 ppm | *Stage II* ⩾0.35 ppm | *Annual basin max*/ppm |
|---|---|---|---|---|---|---|
| 1976 | 237 | 194 | 166 | 102 | 7 | 0.38 |
| 1977 | 242 | 208 | 184 | 121 | 11 | 0.39 |
| 1978 | 217 | 187 | 173 | 116 | 23 | 0.43 |
| 1979 | 226 | 191 | 169 | 120 | 17 | 0.45 |
| 1980 | 210 | 167 | 152 | 101 | 15 | 0.41 |
| 1981 | 222 | 180 | 159 | 99 | 5 | 0.37 |
| 1982 | 191 | 149 | 121 | 63 | 2 | 0.40 |
| 1983 | 190 | 152 | 138 | 84 | 3 | 0.39 |
| 1984 | 207 | 173 | 146 | 97 | 0 | 0.34 |
| 1985 | 206 | 158 | 136 | 83 | 7 | 0.39 |
| 1986 | 217 | 164 | 140 | 79 | 1 | 0.35 |
| 1987 | 196 | 160 | 130 | 66 | 0 | 0.33 |
| 1988 | 216 | 178 | 144 | 77 | 1 | 0.35 |
| 1989 | 211 | 157 | 120 | 54 | 0 | 0.34 |
| 1990 | 184 | 130 | 107 | 41 | 0 | 0.33 |
| 1991 | 183 | 130 | 100 | 47 | 0 | 0.32 |
| 1992 | 191 | 143 | 109 | 41 | 0 | 0.30 |
| 1993 | 186 | 124 | 92 | 24 | 0 | 0.28 |
| 1994 | 165 | 118 | 96 | 23 | 0 | 0.30 |
| 1995 | 154 | 98 | 59 | 14 | 0 | 0.26 |
| 1996** | 152 | 90 | 53 | 7 | 0 | 0.24 |

*The total number of days during the year in which one or more air monitoring stations in the South Coast Air Basin recorded a maximum 1-hour average exceeding the specified level.
**Data for 1996 are based in part on unvalidated data (Oct.–Dec.) and are subject to revision.
Source: SCAQMD, 1997.

(ROG), 50% of the oxides of nitrogen ($NO_x$) and 17% of the carbon monoxide (CO).[5] Thus cars and trucks produce over half of the smog-forming pollution in the state. Although new cars emit 80% fewer pollutants than those produced in the 1970s, a large increase in population and changes in driving patterns have offset the benefits of reduced emissions. The recent increase in the use of sport utility vehicles, which typically pollute more than automobiles, has also tended to offset the improvements otherwise seen in the light-duty fleet. Uncontrolled off-road mobile sources also contribute significantly to emissions. They account for approximately 10% of the ROG, 28% of $NO_x$ and about 17% of the CO.[5] These sources include utility equipment (*e.g.* lawn mowers, riding mowers, string trimmers, blowers, chainsaws and hedge trimmers), off-road motorcycles, construction and farm equipment, locomotives and marine vessels. While these sources may seem insignificant, CARB has indicated that a single chainsaw emits more smog-forming hydrocarbons in 30 minutes than 24 1991 automobiles.[5] As

[5] CARB, *Status Report—1994*, California Air Resources Board, Sacramento, December 1994.

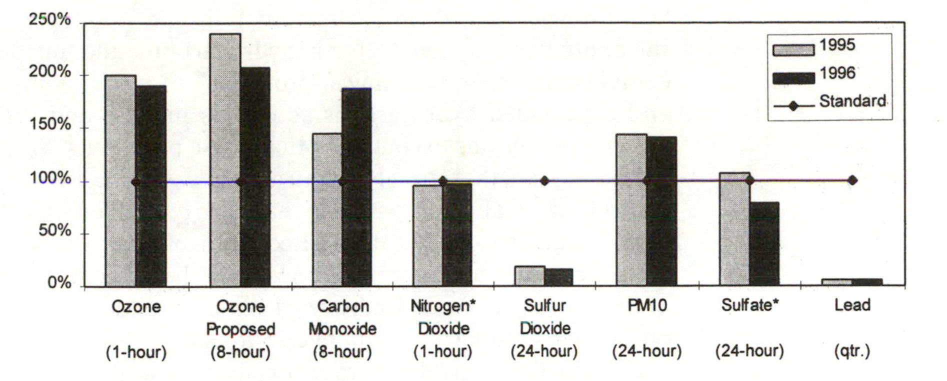


**Figure 4** The 1996 maximum pollutant concentrations as a percent of federal standards. Based on January–December data (October–December unvalidated) for $O_3$, CO and $NO_2$, and January–November data for lead and sulfate.

discussed below, the historical trend has been to control more and more sources, even including small consumer products, because of their contribution to pollutant burdens. As the more obvious, larger sources come under control, the relative importance of smaller but multi-point sources becomes highlighted and is now subject to control.[6]

## 3 Technical Basis and Approach to Emission Controls

California has followed an emission control policy based on sound science, much of which has resulted from its own research programs. The CARB works in conjunction with the US EPA but has pioneered many studies of its own. Its research division, created in 1976, has contributed substantially to the scientific literature on air pollution effects and control requirements. CARB-sponsored studies have investigated the impacts of air pollution on health and its programs have addressed many issues, including the identification of the major pollutants and their sources, the best means of emissions control for California and the characterization of statewide exposure to air toxics. The CARB is currently in the middle of a 10-year multi-million dollar study to assess the health damage to southern California children resulting from continued exposure to ozone, nitrogen dioxide, particulate matter and atmospheric acidity. Results from this study, one of the most extensive of its kind, will in turn help direct the design of control measures for California. In recent years, CARB information has been supplemented by information gained at the SCAQMD, which has a more limited and focused program of research and technology advancement.[7,8]

### *Criteria Pollutants*

Research carried out under the auspices of the CARB and also the EPA has shown the significant role of the criteria pollutants (pollutants for which there are

[6] J. M. Lents and W. J. Kelly, *Sci. Am.*, 1993, **269**, 32.

[7] SCAQMD, *Air Quality Management Plan*, South Coast Air Quality Management District, Diamond Bar, CA, 1996.

[8] SCAQMD, *Annual Report*, Technology Advancement Office, South Coast Air Quality Management District, Diamond Bar, CA, 1996.

health-related air quality standards) in air pollution. Recognition of the substantial interaction between hydrocarbons and nitrogen oxides, through complex atmospheric chemical processes[9] to form ozone, peroxyacetal nitrate and oxygenated hydrocarbons as well as nitrates and sulfates, has led to the design of programs to control both of these pollutants. While it has been argued for many years that stringent control of $NO_x$ can be counterproductive to ozone control,[10,11] California's strategy has clearly been effective in reducing ambient ozone (Figures 1–3 and Table 1). Reduction of sulfur dioxide and lead exposure is far less complex and more easily accomplished by controlling these compounds at the source, *i.e.* use of lead-free gasoline or reduction of sulfur in various fuels burned by stationary and mobile sources.

Historically, control of particulate matter has also been based on reduction at the source, such as controlling fugitive dust from roadways, reducing activities at construction sites, *etc.*[12,13] The expected promulgation[14] of a fine particle standard at 2.5 microns, however, requires an increased understanding of the formation of these particles, which are typical of those which produce photochemical air pollution. As a consequence, control strategies for fine particles are targeted at different sources than those for larger particles. The new $PM_{2.5}$ standards being proposed by EPA are $15\,\mu g\,m^{-3}$ and $65\,\mu g\,m^{-3}$ for the annual and 24-hour averaging period, respectively. The debate about the need for and form of the new fine particle standard is ongoing.[15]

## *Air Toxics*

California defines a toxic air contaminant as a hazardous substance known to cause chronic, often irreversible, health problems such as cancer. Exposure to air toxics is being reduced in a systematic manner by a process in which the highest risk toxic contaminants are identified and included in the risk assessment program. Further measures are then developed under a risk management program carried out by the CARB in cooperation with local air pollution districts. Initially, CARB developed a short list of so-called toxic air contaminants (TAC), but the list was expanded in 1993 to almost 200 substances and includes air pollutants identified under the federal Clean Air Act Amendments of 1990. The CARB process of identifying TACs has been relatively slow and, in some cases, has taken a significant amount of time. For example, the evaluation of diesel exhaust as a potential toxic air contaminant began three or four years ago and has yet to culminate in a recommendation. While this is partly due to the need to reconcile differences in interpreting the technical data, the process has

[9] B.J. Finlayson-Pitts and J.N. Pitts, Jr., *Atmospheric Chemistry: Fundamentals and Experiment Techniques*, Wiley, New York, 1986.

[10] NRC, *Rethinking the Ozone Problem in Urban and Regional Air Pollution*, National Research Council, National Academy Press, Washington, 1991.

[11] J.G. Calvert, J.B. Heywood, R.F. Sawyer and J.H. Seinfeld, *Science*, 1993, **261**, 37.

[12] J.C. Chow, *JAWMA*, 1995, **45**, 320.

[13] J.C. Chow and R.T. Egami, *California Regional Particulate Air Quality Study*, California Air Resources Board, Sacramento, 1997.

[14] M. Nichols, *EM Mag.*, Feb. 1997, 14.

[15] 'Debate on Nature and Need for New Standards,' *EM Mag.*, 1997, in press.

been slowed because of the political implications of controlling diesel exhaust.

A further element of the TAC control programs is the 'hot spots' program, which requires facilities that produce TACs to prepare an inventory of these substances, to make these inventories public and, if necessary to reduce any significant health risk to the local community. One of the benefits of the toxics hot spots program has been the voluntary reduction of toxic pollutants by businesses because of the potential liability resulting from exposure of the broader community beyond their 'fence line'. Many businesses are adopting a pollution prevention approach by substituting for toxic substances other, more desirable, ones, such as the use of citric acid[6] to clean electronic components instead of perchlorethylene or 'Perc'.

The above discussion on TACs addresses emissions from stationary sources. Toxic emissions from the mobile sector have been addressed through the reduction of toxic contaminants in the fuel. For example, ambient benzene concentrations have been reduced substantially as a result of the move to cleaner-burning fuels and to lower aromatic fuels as well as reductions from stationary sources. Figure 2 shows how successful this effort has been.

### *Consumer Products*

As emissions from major sources are reduced, emissions from smaller sources become more significant. Consumer products have been identified as potentially releasing a significant amount of emissions to the atmosphere. For example, it was found that volatile organic compounds (VOCs) released from consumer products total up to 200 tonnes per day statewide.[5] Thus, regulations were adopted in 1989 and 1990 for antiperspirants and deodorants and additional products came under control in 1992. Table 2 lists products subject to regulation under the CARB program as of 1994. It is anticipated that emissions from current consumer products will be cut by about 30% by the year 1999.

### *Stationary Sources*

In addition to the control of toxic emission from other stationary sources, a successful program has been the capture of vapors at gas stations. So-called vapor recovery systems, while clumsy in earlier years, have improved substantially and now are very efficient at reducing the emissions of VOCs including TACs to the atmosphere. For example, they substantially reduce exposure to TACs of people refueling their cars at gas stations.

## 4 Philosophy of Air Quality Management

Long-range plans to address air pollution, required under both state and federal laws, come together through State Implementation Plans (SIP). These must be prepared by all states in the US. California's state implementation plan addresses all sources of smog-forming and particulate emissions. It covers a 15-year time frame and integrates state, local and federal components, providing a comprehensive

**Table 2** Regulated consumer products (CARB, 1994)

| 1989 *Regulations* | | |
|---|---|---|
| Antiperspirants and deodorants | General purpose cleaners | Laundry prewash materials |
| Air fresheners | Glass cleaners | Oven cleaner |
| Bathroom and tile cleaners | Hairspray | Nail polish removers |
| Engine degreaser | Hair styling gels | Shaving creams |
| Floor polish | Hair mousse | Windshield washer fluid |
| Furniture maintenance materials | Insect repellents (aerosols) | |
| *Added in* 1992 | | |
| Aerosol cooking sprays | Dusting aids | Laundry starch |
| Brake cleaners | Fabric protectants | Perfumes and colognes |
| Carburetor choke cleaners | Household adhesives | Insecticides |
| Charcoal lighter fluid | | |

strategy.[16] Elements of the plan are submitted by the various local districts to the CARB, which then approves those plans and, in turn, submits an overall statewide SIP to the US EPA. This is reviewed by the EPA and, when approved, commits the state to the actions contained in the plan. In recent years, as it has become more difficult to identify specific control measures, various reduction components have been put into a so-called 'black box'—the specific controls to effect the emissions reductions will be identified in future years as new technologies are identified.

In 1988 the California Legislature passed the California Clean Air Act (CCAA). This was seen as another example of the strong commitment of California's leaders to provide additional impetus to cleaning up the air. The CCAA included provisions to accelerate CARB's regulation of vehicles and fuels. It also extended the regulatory scope of local districts and CARB to include off-highway vehicles, locomotives, marine vessels, farm and construction equipment, utility engines, transportation sources and traffic generating facilities, and development projects. While several of these provisions, including the control of locomotives and marine vessels were eliminated by the federal Clean Air Act Amendments of 1990 (which allowed only federal control of these sources), the California act did stimulate development of new regulations for cleaner vehicles and fuels and resulted in, among others, the low emissions vehicle programs in California, described in more detail in Section 7.

Control of pollution can be approached in two general ways: by a 'command and control' approach or a market incentive program. In California, command and control was the preferred method until the early 1990s, when economic factors played a significant role in shaping environmental policy in the state. The

[16] CARB, *California State Implementation Plan*, California Air Resources Board, Sacramento, CA, 1994.

substantial success that California saw in reducing mobile source emissions was through the promulgation of landmark regulations which forced new technologies. The list of such regulations is lengthy but key mobile source-related examples are:

- the introduction of the catalytic converter
- the phase-out of lead in gasoline
- the reduction of sulfur in fuels
- more recently, the low emission vehicle (LEV) program, which stimulated a dramatic improvement in emission control technology for automobiles

For stationary sources, similar programs have been effective in forcing new technologies to reduce the emissions of criteria and toxic pollutants. For example:

- the requirement to use less reactive solvents
- the reduction of the volatility of fuels
- the capture of vapors at gasoline dispensing stations, through vapor recovery systems
- replacement of oil and coal by natural gas for power generation
- reduction of the sulfur content of fuels

Market incentive programs rely on economic incentives to encourage polluting entities to reduce emissions. This philosophy is now much in vogue not only in California but throughout the US. It is based on the principle of setting an emissions target for industry and allowing the least-cost economic approach to reduce emissions.

The most extensive and well recognized market based program is the RECLAIM (Regional Clean Air Incentives Market) program in the SCAQMD.[17] Under this program a business is asked to meet emission limits not only for an individual piece of equipment but also for its entire operations and facilities. Businesses that cut pollution below the standard will be granted emission reduction credits. Companies that generate more pollution than the standard must buy enough credits to make up the difference. In theory the program not only gives firms the flexibility to choose how to reduce emissions but it also provides incentives for innovation—firms can make a profit by cleaning up the air.[18] This is one way in which the market may give rise to technological innovation. Verification and enforcement issues remain with this approach and only time will tell whether it will result in accomplishments as significant as 'command and control', which have been documented in the earlier discussion.

This change in philosophy regarding emissions reductions during the 1990s is discussed further in Section 8. The economic recession within the state has influenced the vigor with which control strategies have been pursued recently. This is particularly true of the SCAQMD, where plans to reduce emissions have been slowed significantly through delays in implementing regulations and in the delay in dates on which rules would take effect. The number of citations issued to

[17] SCAQMD, *Regional Clean Air Incentives Market (RECLAIM), Revised Program Development Report*, South Cost Air Quality Management District, Diamond Bar, CA, 1993, Vols. I–V.

[18] C. Moore and A. Miller, *Green Gold: Japan, the United States, and the Race for Environmental Technology*, Beacon Press, Boston, 1994.

industries for violating their operating permit conditions has also been reduced. This is in sharp contrast to the aggressive programs in the 1988–93 time period. At the state level, further similar slowing of movement has taken place; one can see the significant impact of economic considerations in the CARB's 1995 Mission Statement, which read:

> 'To promote and protect public health, welfare and ecological resources through the effective and efficient reduction of air pollutants while recognizing and considering the effects on the economy of the state.'

Prior to this time, the state's economic health was not given such explicit consideration.

## 5 On-road Vehicle Emissions

Because on-road motor vehicle emissions represent the largest component of the emissions inventory in California, it is important that this information is known with reasonable reliability. The data in the inventory are utilized in computer models to predict current and future control strategies to reduce exposure to the criteria pollutants. Major progress has been made in reducing on-road vehicle emissions. As stated above, this is due to stringent emission control standards introduced in California and made possible as a result of the catalytic converter and subsequent improvements in fuels and other emission controls. However, it was found[19] that model predictions were significantly underestimating actual on-road emissions of hydrocarbons and carbon monoxide. The work carried out by Pierson *et al.*, based upon experiments carried out in traffic tunnels, indicated a major discrepancy between predicted and measured emissions of these pollutants. Subsequent work[20,21] has confirmed that on-road emissions are higher than predicted by the models. These conclusions are also confirmed by remote sensing studies[22,23] which highlight the 'richer' operation of vehicles on the road than predicted by the models.

The CARB is addressing these issues through a combination of improved model development for more accurately predicting on-road emissions, improved inspection and maintenance (I/M) programs, and the use of remote sensing to identify some of the high emitters. This represents a significant change in direction; originally CARB was skeptical of data which showed significant under-prediction by the models. In addition, CARB is expecting significant improvement of on-road emissions performance measurements through the introduction of the on-board diagnostic program (OBD II) in 1996 and new vehicles. This new tool, which monitors the operation of the motor vehicle, is expected to lead to

[19] W. R. Pierson, A. W. Gertler and R. L. Bradow, *J. Air Waste Manage. Assoc.*, 1990, **40**, 1495.

[20] N. F. Robinson, W. R. Pierson, A. W. Gertler and J. C. Sagebiel, *Atmos. Environ.*, 1996, **30**, 2257.

[21] W. R. Pierson, A. W. Gertler, N. F. Robinson, J. C. Sagebiel, B. Zielinska, G. A. Bishop, D. H. Stedman, R. B. Zweidinger and W. D. Ray, *Atmos. Environ.*, 1996, **30**, 2233.

[22] R. Slott, *Proceedings of the Sixth CRC On-Road Vehicle Emissions Workshop*, San Diego, CA, March 18–20, 1996, Coordinating Research Council, Atlanta, pp. 8–9.

[23] R. Goodwin and M. Ross, *Proceedings of the Sixth CRC On-Road Vehicle Emissions Workshop*, San Diego, March 18–20, 1996, Coordinating Research Council, Atlanta, GA, pp. 4–115.

**Table 3** Standards for heavy-duty vehicles for the year 2004

| Manufacturers would have the flexibility of choosing between two options: |
|---|
| • a combined NMHC + $NO_x$ standard of 2.4 g $bhp^{-1} h^{-1}$, or |
| • a combined NMHC + $NO_x$ standard of 2.5 g $bhp^{-1} h^{-1}$ and a NMHC cap of 0.5 g $bhp^{-1} h^{-1}$ |
| These standards are to be achieved while maintaining the 0.1 g $bhp^{-1} h^{-1}$ PM standard established in 1994 |

substantial additional reductions in both hydrocarbon and CO emissions. However, past experience indicates that the effectiveness of the approach utilized in the OBD II to control emissions will need to be clearly monitored by comparing actual on-road performance with the emissions models. The effectiveness of I/M programs in improving on-road vehicle emissions has been questioned.[24] The role of centralized *vs.* decentralized testing has been decided on largely political rather than technical grounds.

As emissions from the light-duty fleet come under greater control, the relative contributions from the heavy-duty fleet become more significant, particularly for $NO_x$ and particulates.[25] Emissions from heavy-duty vehicles have been reduced but not as dramatically as for the light-duty fleet. Programs to identify 'smoking' vehicles have reduced the number of vehicles with highly visible emissions. As a result of a statement of principles (SOP)[26] among the CARB, EPA and the engine manufacturers, new standards shown in Table 3 have been developed for introduction in 2004. While technically challenging, these do not appear to represent as significant a hurdle as that overcome by light-duty vehicles. However, the SOP also calls on the signatories to pursue a research agreement with goals of 1.0 g $bhp^{-1} h^{-1}$ $NO_x$ and 0.5 g $bhp^{-1} h^{-1}$ particulate matter while maintaining current diesel engine performance. A similar agreement was signed by CARB and EPA with major manufacturers of diesel-powered farm and construction engines and equipment.[27]

In addition to the technological solutions programs to reduce on-road emissions, a number of transportation measures were developed by local agencies, including car pooling, the use of high-occupancy vehicle lanes in traffic, telecommuting and improved public transit. These programs have had limited success and their effects are difficult to quantify. The reaction from the business community to the ride-share program in SCAQMD continues to be negative and led to a significant backlash in the California Legislature, which supported industry efforts to eliminate this regulation. The major issues were the amount of money that industry was required to spend and the uncertain air quality benefits from the regulation.

The CARB has also studied various aspects of transportation-related land use strategies, including indirect sources such as shopping centres.[28,29]

[24] W. R. Pierson, *Atmos. Environ.*, 1996, **30**(21), i–iii.

[25] SCAQMD, *Air Quality Management Plan*, South Coast Air Quality Management District, Diamond Bar, CA, 1996, Appendix IV–A.

[26] US Environmental Protection Agency, Washington, Report No. EPA-A420-F-95-008a.

[27] California Air Resources Board, Sacramento, New Release 96-23, September 19, 1996.

[28] California Air Resources Board, Sacramento, Research Notes, No. 95-18, September, 1996.

[29] California Air Resources Board, Sacramento, Research Notes, No. 95-27, December, 1995.

## 6 Cleaner Burning Fuels—Rationale and Implementation

California instituted efforts to change the composition of gasoline with the introduction of the catalytic converter in 1974. Converters are poisoned by lead in gasoline; hence unleaded gasoline was required. Subsequently it was found that emissions of toxic air contaminants such as benzene increased because oil companies, to maintain octane rating, made up for the absence of tetraethyllead by increasing the aromatic content of the fuel. This led to efforts to reduce the aromatic content as well as the sulfur content in fuels, the latter to avoid poisoning the converter.

In the early 1980s the desire to achieve even lower emissions led to renewed interest in alternative fuels including methanol, natural gas, propane and ethanol. During the mid-1980s, California air pollution agencies as well as the California Energy Commission (CEC) became increasingly interested in methanol as a fuel for both mobile and stationary sources. The goal was to provide a cleaner burning fuel which would lead to lower automobile emissions. Under California Legislation AB-234, the state conducted an evaluation of the need for clean fuels and their impact on air quality and highlighted the potential for alternative fuels to reduce emissions.[30] At the same time, given the threat of methanol sales undercutting gasoline profits, the Atlantic Richfield Corporation began producing a cleaner burning gasoline, EC-1. This first effort to provide cleaner burning gasoline has most recently led to the deployment of reformulated gasoline in California. This reformulated gasoline (RFG) has lower Reid Vapor Pressure (RVP), which has reduced evaporative emissions. In addition, RFG has reduced aromatic, olefinic and sulfur content compared with conventional gasolines. (Sulfur content continues to dictate the lowest levels to which emissions may be reduced because it influences the ability of catalytic converters to reduce emissions.)

The results of introducing alternative fuels have been mixed in spite of significant encouragement by the CARB, CEC and the SCAQMD. The latter created a Technology Advancement Office in 1988 to assist the automotive and fuel industries in expediting development and implementation of new pollution control technologies and clean burning fuels. To date, nearly $50 million has been contributed by the agency to programs which total over $250 million. The major methanol and natural gas introduction programs have been stalled by lack of infrastructure. The electric vehicle program is just now developing in a serious way (see below). Use of alternative fuels in the heavy-duty fleet has also had mixed results. Buses operating on natural gas in Sacramento, California, have had reduced emissions and operating costs and have been judged a success. The major bus fleet in Los Angeles has had mixed results with both natural gas and methanol and they are currently experimenting with a number of ethanol-powered buses. The electric bus fleet in Santa Barbara has operated successfully for a number of years.

A barrier to the greater use of these alternative fuels is the continued reduction in emissions from vehicles operating on cleaner gasolines and cleaner diesel fuel. The cleaner burning gasolines, together with improvements in vehicle emissions technology, have been greatly responsible for the substantial improvement in air

[30] California Assembly Bill AB-234, *Impact of Alternative Fuels on Air Quality*, 1988.

quality in California, particularly in southern California. Table 1, discussed earlier, illustrates the improvement in ozone during the period 1976–96. It is difficult to attribute these results to one strategy. However, it seems very clear that the recent introduction of cleaner burning gasoline, which affects the total on-road fleet, has led to significant benefits in a short time. New technologies which rely on replacement of the on-road vehicle fleet is a much slower process.

Cleaner diesel fuel has had limited success and the introduction of reformulated diesel a few years ago in California led to significant operational problems for trucks. It is not clear whether these problems are related to an increase in fuel costs or to real technical problems resulting from damage to engine seals, *etc.*, related to the new fuel. As a result, there has been a warier approach to introducing different fuels into commerce. For example, the introduction of RFG in June 1996 was managed more carefully; government agencies and oil companies worked together to ensure that the public was minimally affected. The increase in the price of the fuel (around 10–15 cents per gallon) did not meet significant public disapproval, largely because of an effective educational program ahead of time.

One problem which has arisen recently is the contamination of certain reservoirs and lakes by methyl *tert*-butyl ether (MTBE). This additive boosts the oxygen content of gasoline, thus increasing octane rating. Benefits of providing cleaner burning gasoline may result in water quality problems because MTBE is miscible with water. MTBE has been detected in lakes and reservoirs utilized by jet skis and speed boats; the MTBE originates from gasoline leakage into the water bodies. This exemplifies the need to examine the impacts of a particular strategy on the total environmental system to ensure that improvement in one area does not result in degradation of another.

## 7 Low Emission Vehicle Program

Although there have been substantial improvements in emissions from motor vehicles, the mobile source component of the emissions inventory remains the largest in California. To provide additional stimulus to reduce emissions from mobile sources, the CARB adopted the Low Emission Vehicles (LEV) program in 1990.[31] This program requires vehicle manufacturers to phase in progressively cleaner operating light- and medium-duty vehicles, culminating in the introduction of zero emission vehicles. For the first time, an increasingly stringent annual fleet average emission requirement was established to provide a flexible mechanism for phasing in lower emission vehicles. Table 4 shows the emission standards for each vehicle class: transitional low emission vehicles (TLEVs), low emission vehicles (LEVs), ultra-low emission vehicles (ULEVs) and zero emission vehicles (ZEVs). The fleet averaging requirements are shown in Table 5.

Originally, 2% of new vehicle sales in 1998 were to be zero emission vehicles; this percentage would increase to 10% by the year 2003. The 1990 regulations required CARB to conduct a biannual technology review. In 1996 it became apparent that the LEV program needed modification. Specifically, it was found in

[31] CARB, *Proposed Amendments to Low-Emission Vehicle Regulations*, California Air Resources Board, Mobile Source Division, El Monte, CA, 1995.

**Table 4** Light-duty low-emission vehicle exhaust emission standards

| *Vehicle class* | *NMOG** | *CO* | $NO_x$ |
|---|---|---|---|
| Tier† | 0.25 | 3.4 | 0.4 |
| TLEV | 0.125 | 3.4 | 0.4 |
| LEV | 0.075 | 3.4 | 0.2 |
| ULEV | 0.040 | 1.7 | 0.2 |
| ZEV | 0 | 0 | 0 |

*'NMOG' is non-methane organic gas and is composed of non-methane hydrocarbons and all oxygenated hydrocarbons.
†'Tier 1' refers to the non-methane hydrocarbon (NMHC) standard which applies to conventional gasoline vehicles.

**Table 5** Fleet average requirements for passenger cars and light-duty trucks (0–3750 lbs)

| *Model year* | 1994 | 1995 | 1996 | 1997 | 1998 | 1999 | 2000 | 2001 | 2002 | 2003 |
|---|---|---|---|---|---|---|---|---|---|---|
| Fleet average NMOG | 0.250 | 0.231 | 0.225 | 0.202 | 0.157 | 0.113 | 0.073 | 0.070 | 0.068 | 0.062 |

the ZEV component that the available battery technology (typically lead acid batteries) provided too short a driving range; the advanced battery likely to increase performance significantly would not be available until around the year 2001. The most promising battery appeared to be the lithium family of batteries with intermediate improvements in range coming from nickel/metal hydride and nickel/metal chloride batteries. It was also felt that a more gradual, voluntary introduction of ZEVs by the auto companies would be more desirable, resulting in 'memoranda of agreement' between CARB and the auto companies.[32] Thus, the sales percentage requirements for 1998 to 2002 were eliminated, but the 10% requirement for 2003 was retained. One consequence of this delay is the greater chance that fuel cell powered vehicles[33] can play a role in meeting the 2003 requirement and help offset some of the battery technology limitations. Technology for fuel cell vehicles (FCV) has improved substantially in recent years,[34] and prototype vehicles have been marketed by Daimler-Benz, Toyota and Chrysler.

Another important component of the 1990 LEV regulations was the attempt for the first time to account for the difference in reactivity of the exhaust emissions. Inclusion of this concern in a regulatory framework was recognition that different hydrocarbons react in the atmosphere with nitrogen oxides and sunlight to different extents to form ozone. The LEV program incorporated reactivity adjustment factors (RAFs) to account for this difference.

The reactivity adjustment factors developed to date (shown in Table 6) have generated a significant amount of discussion. However, they provide explicit recognition that, while different fuels may emit different quantities of hydrocarbons,

[32] CARB, *Memoranda of Agreement between CARB and Various Auto Companies*, California Air Resources Board, Mobile Source Control Division, El Monte, CA, 1996.
[33] A.C. Lloyd, J.H. Leonard and R. George, *J. Power Sources*, 1994, **49**, 209.
[34] J. Tachtler and C. Bourne, *Fuel Cell: Program and Abstracts*, Proceedings of the Fuel Cell Seminar, November 17–20, 1996, Orlando, FL, pp. 266–269.

**Table 6** Reactivity adjustment factors

| | *Light-duty vehicles* | | | *Medium-duty vehicles* | |
|---|---|---|---|---|---|
| | *TLEV* | *LEV* | *ULEV* | *LEV* | *ULEV* |
| *Fuel* | *Baseline specific reactivity* ($g\,O_3/g$ NMOG) | | | | |
| Conventional gasoline | 3.42 | 3.13 | 3.13 | 3.13 | 3.13 |
| | *RAFs* | | | | |
| Phase 2 RFG | 0.98 | 0.94 | 0.94 | 0.94 | 0.94 |
| M85 | 0.41 | 0.41 | 0.41 | 0.41 | 0.41 |
| Natural Gas | 1.0 | 0.43 | 0.43 | 0.43 | 0.43 |
| LPG | 1.0 | 0.50 | 0.50 | 0.50 | 0.50 |
| E85* | — | — | — | — | — |

Source: ARB Staff Report: Initial Statement of Rulemaking, August, 1995.
*No data.

the most important consideration is the ability of the exhaust to contribute to the ozone formation. The overall influence on ozone formation is determined by multiplying the RAF by the total mass NMOG emissions from the exhaust.

## 8 Economy of California and its Influence on Air Quality Management in the State

During the early 1990s, California suffered a significant and sustained economic downturn. Although this was largely due in southern California to a cutback in funding for military aerospace applications, many people used this as an opportunity to blame environmental regulations for the economic problems. Strict environmental regulations were blamed for driving business out of the state. Pressure extended to the California legislature. Less than a decade after they had encouraged the SCAQMD (the largest air district in the state) to be more aggressive in cleaning up the air, the legislature put significant constraints on the District's ability to carry out its legal charter—budget reductions and limitations on the District's ability to raise money were effectively enforced through legislation in Sacramento. The legislature also required that the CARB review the SCAQMD annual budget forecast.[35] In addition, air pollution control at the state level was also softened, although not as significantly as at some of the local levels. As a result, the level of enforcement and the number of new regulations have been significantly reduced from the late 1980s. Many of the new regulations rely more on market-based incentives and on self-policing activities. Concern about the easing of the regulatory approach resulted in the resignation of nine of the eleven scientific advisors to the SCAQMD in the summer of 1996.[36]

Were the concerns of the Legislature and industry valid? The impacts of the strict environmental regulations on the state's economy were studied by Hall *et*

[35] California Assembly Bill Ab-1853 (Polanco), September 1994.
[36] M. Cone, *Los Angeles Times*, 1996, August 9, 1.

*al.*[37] (California State University at Fullerton). They found no evidence that the strict regulations had a major impact on the recession. Many factors played a role; it could also be argued that some industries profited by meeting California's strict environmental regulations, which allowed them to market their equipment and processes both nationally and worldwide.[38]

## 9 Summary

Air quality in California has shown significant improvement despite large increases in population and vehicle miles traveled. This success can be attributed to many years of consistent and tough emission standards on both mobile and stationary sources coupled with strong enforcement programs.

The LEV program, which addresses both vehicle emissions and fuels, has made a significant contribution, not only to cleaning the air, but in stimulating technological innovation both nationally and internationally. Transportation control measures have offered limited success.

While air quality has improved significantly, the health-related air quality standards are still exceeded on a regular basis. The new standards for ozone and $PM_{2.5}$ are more difficult to meet than existing standards. Major emphasis has now been placed on the market incentive approach through public–private partnerships. Whether this approach will continue the substantial progress demonstrated under 'command and control' remains to be seen.

## 10 Acknowledgements

The author wishes to thank the staff of the California Air Resources Board and the South Coast Air Quality Management District for providing data for this article. Thanks are also expressed to Beverly Finley and Susan Grobman for their significant efforts in the preparation of this paper.

[37] J. Hall, J. A. Aplet, S. Levy, G. Meade and A. Puri, *The Automobile, Air Pollution Regulation and the Economy of Southern California, 1965–1990*, prepared for W. Alton Jones Foundation, Charlottesville, WV, 1995.

[38] C. Moore and A. Miller, *Green Gold – Japan, Germany and the United States and the Race for Environmental Technology*, Beacon Press, Boston, 1994.

# Subject Index